ABSINTHE &
FLAMETHROWERS

ABSINTHE &
FLAMETHROWERS

PROJECTS AND
RUMINATIONS
ON THE
ART OF LIVING
DANGEROUSLY

WILLIAM GURSTELLE

CHICAGO
REVIEW
PRESS

Library of Congress Cataloging-in-Publication Data

Gurstelle, William.
  Absinthe & flamethrowers : projects and ruminations on the art of
living dangerously / by William Gurstelle. — 1st ed.
    p. cm.
  Includes bibliographical references and index.
  ISBN 978-1-55652-822-4
  1. Threat (Psychology) 2. Daredevils. 3. Stunt performers. I. Title.

BF575.T45G87 2009
155.2'32—dc22

                                            2008046619

Visit www.AbsintheAndFlamethrowers.com for additional help and updates.

Cover design: Joan Sommers Design
Interior design: Rattray Design

First edition
Published by Chicago Review Press, Incorporated
814 North Franklin Street
Chicago, Illinois 60610
ISBN 978-1-55652-822-4
Printed in the United States of America
5 4 3 2 1

Acknowledgments

The information in this book has been assiduously researched and documented. In addition to a tremendous amount of library and database research, I interviewed some of the world's leading experts on what are unquestionably some very arcane and esoteric subjects.

I interviewed several academic researchers whose pioneering work has opened new areas of psychological research within the topics of risk and risk taking. What a thrill it was to learn from the person who knows so much about fast cars and fast driving that he's the one Jay Leno turns to for help. I talked extensively with the pyrotechnical engineer who literally wrote the book on making black powder. I spoke at length with a nationally famous chef and television show host who travels the world eating dangerous foods.

The august group of learned academicians, tech wizards, and practical geniuses—my panel of experts—patiently answered questions, provided key pieces of information, and helped me form the ideas and projects contained in this book.

All were willing to take time and share their expertise with me. I thank them with great sincerity.

The Living Dangerously Panel of Experts:

Alan Abel—Writer and Media Prankster
Gale Banks—Automotive Expert, Founder of Gale Banks
    Engineering
Robert Dante—Bullwhip Artist
Dave DeWitt—Author and Chile Pepper Consultant

Dr. Frank Farley—Professor of Educational Psychology, Temple University

Dr. Alain Goriely—Professor of Mathematics, University of Arizona

Mr. Jalopy—Contributing Editor, *Make* magazine

Morten Kjølberg—Founder, Lightertricks.com

Dr. Dirk Lachenmeier—Food Scientist, CVUA Karlsruhe, Germany

Dr. Stephen Lyng—Professor of Sociology and Criminal Justice, Carthage College

Dr. Terry McCreary—Professor of Chemistry, Murray State University

Richard Nakka—Rocket Engineer

Christian Ristow—Mechanical Artist

Jon Sarriugarte—Owner, Form and Reform

Ian von Maltitz—Author and Engineer

Andrew Zimmern—Host, Travel Channel's *Bizarre Foods*

Dr. Marvin Zuckerman—Emeritus Professor of Psychology, University of Delaware

Besides the panel of experts, there are others to whom I owe a great deal. I am much obliged to the support and inspiration provided by my family: Alice Gurstelle, Carol and Steven Gurstelle, Carol and Glen Fuerstneau, and Wendy and Barry Jaffe. My agent, Jane Dystel at Dystel and Goderich, has long been a source of help and a pillar of consistency in a rapidly changing publishing world. Cynthia Sherry, publisher and editor at Chicago Review Press, made many significant and thoughtful suggestions for which I am very grateful. Conversations regarding this work with my son (and attorney-in-training) Ben and my son (and personal archeologist) Andy were always helpful and uplifting. There are others who helped and inspired me but are not listed here due either to my oversight or their understandable desire to keep their name out of the papers. Finally, I owe a tremendous debt of gratitude to Karen Hansen, who helped me in so many ways as I wrote this book, just as she does in all of the other aspects of my life.

Contents

The Age of the Lily-Livered

*We are living in the age of the lily-livered, where everything is a pallid
parody of itself, from salt-free pretzels to the schooling of children amid
foam corner protectors and flame-retardant paper.*

I blame the people at the top for setting the tone.

I blame parents.

I blame the arbiters of virtue.

*Sometime over the past generation we became less likely to object to
something because it is immoral and more likely to object to something
because it is unhealthy or unsafe. So smoking is now a worse evil than six
of the Ten Commandments, and the word sinful is most commonly associated with chocolate.*

*Gone, at least among the responsible professional class, is the exuber-
ance of the feast. Gone is the grand and pointless gesture.*

—ABRIDGED FROM AN ESSAY BY
DAVID BROOKS, *NEW YORK TIMES*,
MARCH 12, 2005

I admit to being somewhat of a late bloomer in terms of striving
to be a person who lives dangerously and artfully. For two
decades I worked in the communications business, a notoriously
procedure-bound industry, doing presumably important but not
particularly engaging work. This industry is huge, multifaceted,

and full of opportunities. For more than 100 years, engineers like me provided highly dependable telephone service to anyone who could afford the incredibly modest rental price of a phone line.

While some perhaps looked forward to each workday with breathless anticipation, to me it was monotonous and ultimately stultifying. Going to work every day became a drag, and I found myself basically waiting for the weekend.

Sure, salaries were decent, and without question there are opportunities for those who base their careers here to learn and grow. But I never talked about what I did at cocktail parties. It did not make for interesting conversation. What I did was honorable and, I suppose, important, but it was devoid of art and bereft of danger.

Eventually I figured out that the path I had chosen wasn't the right one for me. I realized there was a hole in my life, an emptiness caused by missing the experiences and risk/reward calculus that my brain chemistry required for happiness.

In 1999 I began investigating and writing about unusual, highly kinetic science and technology. I had always been interested in homemade things that went *whoosh* and *boom*, but I hadn't really considered them serious enough or important enough to turn into a career.

The turning point occurred when I wrote a book of science experiments called *Backyard Ballistics*. The unique thing about the dozen science projects in the book is their edginess, their quirkiness, and most of all, their obvious potential for danger. In the book are descriptions of the homebuilt, high-velocity, high-energy experiments and projects I'd been collecting since I was a freshman in college. There are instructions for building things that shoot potatoes at 80 miles per hour, that send kites of burning newspapers high overhead, and that fling projectiles at high speed from the slings of homebuilt catapults.

I was all but certain that no publisher would take a chance on such a book, considering the litigious nature of modern society. But a respectable publisher was interested, in fact more than one. Happily, there are people out there who recognize a need for books and information that appeal to people who want to do and experience interesting things themselves. Now, several years and a quarter million copies later, *Backyard Ballistics* continues to sell briskly.

Since my first book was published in 2001, I've received nearly a thousand e-mails and letters from readers. Some were eager to share their results, some had questions, and some wanted to share ideas for projects that they developed themselves. I read every message, and while I too often may have depended on e-mail autoresponders and FAQ sections on my Web site, I value every reader communication I received.

The messages that I particularly prize are stories of how *Backyard Ballistics* or one of my other books changed some aspect of a person's life. Some wrote to tell me how taking on the projects in the book led to a choice of school, major, or even career. Others told me how doing those projects led them to a closer relationship with a father, mother, son, daughter, or friend. Others stated that doing the projects simply made them interested in science and learning.

The common thread was that they enjoyed trying out projects that seemed to be at least a little dangerous. Not crazy-dangerous, mind you, but not the pap and pabulum that is too often presented and passed off as science exploration nowadays. These readers had been waiting, it seemed, for a book that respected their ability to follow directions, to think for themselves, and to assume responsibility; a book where the information they sought was made clearly available and allowed them to do with it what they would.

Writing these books has been a wonderful experience. In the course of doing so I've perhaps shot more things from the sling of a catapult or the barrel of an air cannon than any person on earth. I've built catapults, tamped rocket motors, milled explosives, assembled taser-powered cannons, and fabricated fighting robots. In the course of writing this book I've added several new experiences, such as stick fighting, fugu eating, and rattlesnake hunting.

I've written scores of magazine articles and I've lectured on these topics at a host of different venues in Asia, Europe, and Australia as well as all over the United States.

Rather than avoid experiences that might be considered risky, I've decided to seek those that make sense to me, to attempt as much edgework (a term I define in chapter 2) as I can find as long as they conform to a certain conceptualization for what is both dangerous *and* artful.

Learning to live dangerously and artfully will likely seem extreme to some and tame to others. That's fine, for I'm happy residing in what I term the Golden Third, which as you'll see later in this book is the statistical area where risk and adventure coexist in rational, relative proportion. For in that wonderful slot, there is no need to change a career or significantly alter your lifestyle to satisfy your brain's requirement for intellectual excitement or danger. You won't need to leave your family or move to a foreign country.

People take risks; they risk money, reputation, and physical well-being for many reasons. Sometimes risk taking is mandatory; no other option presents itself. But sometimes people take risks for no apparent reason and in apparent opposition to fundamental instincts for safety. Are such people acting illogically, against instinct, and therefore against nature?

Recent research findings would indicate not. People fall all over the map in terms of the experiences they seek. What seems logical for you may seem overly dangerous for me. Basically, people are motivated to take risks as a result of their psychological makeup, the situation they find themselves in, and the particular way their brain functions.

Marvin Zuckerman, professor emeritus at the University of Maryland, wrote a magnificently elucidating book on the subject. It explores the nature of human risk taking and should be better known than it is, although it is an academic work burdened with the not-so-snappy title of *Behavioral Expressions and Biosocial Bases of Sensation Seeking*. In it, Zuckerman brilliantly quantifies human affinity for seeking and enjoying new sensory experiences. He developed a personality test that measures affinity for risk. Since he devised it in the 1960s, thousands have taken Zuckerman's test or questionnaire and a huge body of statistical information has been formed, allowing great insights into the sociology of human risk seeking.

The test itself is easy enough to understand: based on answers to a scientifically designed questionnaire, an individual receives a score that describes the degree to which that person seeks sensations and takes risks. That single number can be broken down into four more refined attributes: thrill seeking, experience seeking, disinhibition (sensation seeking through social activities such as parties, social drinking, and sex), and boredom avoidance. In chapter 3, you'll have the opportunity to take a modified version of Zuckerman's questionnaire, hopefully providing you with a starting point for understanding your own risk-taking psychology.

Besides Zuckerman, many others have written about the psychobiology of risk taking. The research indicates there are complicated chemical processes at work in the human brain that push people in one direction or the other in terms of sensation seeking and risk taking. Researchers are able to identify the specific brain chemicals—dopamine, monoamine oxidase, and norepinephrine, among others—that underlie the personality traits of risk taking, impulsivity, and self-preservation.

The details of psychopharmacology and the manner in which brain chemicals cause one person's neurons to fire one way and the next person's another are complicated, the results of a million years of brain evolution. Another Zuckerman work, a book he edited called *Biological Bases of Sensation Seeking, Impulsivity, and Anxiety,* provides a thorough treatment of the relationship between heredity, brain chemistry, and risk-taking behavior. It's not an easy book to understand, involving concepts such as genetics, psychophysiology, and above all, biochemistry.

But what is easy enough to comprehend is that the novelty, sensations, and type of risk-taking experiences you seek, and the extent to which you seek them, are based on your own specific, customized soup of brain chemicals—and that combination of chemicals is a result of the choices made by a thousand generations of your ancestors.

But significantly, you are not condemned by your genetics to a life of either lily-livered risk avoidance or Evel Knievel-like, bone-breaking peril. *You have control and can choose an artful yet exciting path.* The answer is in your hands and what you do with them. That's what this book is about: doing interesting, exciting, edgy, and artful stuff.

Part I

Why Live Dangerously?

1

Big-T People, little-t people

"Life is either a daring adventure or nothing."
—HELEN KELLER

On June 18, 1952, the headline on the front page of the *Los Angeles Times* read "Rocket Scientist Killed in Pasadena Explosion." The unlucky scientist was John Whiteside Parsons, a brilliant but (putting it charitably) strange man who had founded the world-famous Jet Propulsion Laboratory in Pasadena, California. Later he went on to start the now gigantic Aerojet Corporation, a major space contractor specializing in missile and space propulsion whose products include the Atlas, Titan, and Delta rocket engines.

In the early afternoon of what had been a quiet day in this pleasant, well-to-do suburb of Los Angeles, an immense explosion rocked the neighborhood. Heard a mile away, the blast tore apart the aging three-story mansion at 1003 South Orange Grove Avenue. The neighbors had grown accustomed to bizarre goings-on at this address, for there was always a strange mix of people going in and out—bohemian artists, science fiction writers, and occultists to name but some. But this was serious.

Sirens screaming, trucks bearing firemen from all over Los Angeles soon arrived. A few daring individuals chanced the

3

smoking ruins to search for any survivors trapped within. Pushing aside the charred rubble, they found Parsons, at least what was left of him, covered by an overturned washtub. The rescuers gasped when they first turned him over. Several pieces of Parsons, including his arm and the side of his face, were missing. They discerned a weak pulse and frantically dragged him outside, where they hoped and waited for a miracle that never came. His situation was hopeless and he succumbed to his injuries about an hour later.

His cause of death was determined to be an accidental explosion from careless handling of an oversized batch of fulminate of mercury, a highly unstable contact explosive. Producing fulminated mercury requires precise laboratory procedures and precipitation reactions that involve dissolving metallic mercury in nitric acid and adding precise quantities of ethyl alcohol until crystals of the explosive precipitate out of the solution.

Most chemists will tell you that fulminate of mercury is far too dangerous to be made in a home laboratory in anything but the smallest quantities. It is not only poisonous but also so unstable that almost anything will cause it to explode. A small bump, an inadvertent knock, or a short drop is all it takes. Simply jarring it a bit can set it off.

Police conducted a thorough investigation of the scene. They found the remains of numerous containers of different kinds of explosives. Piecing together the forensic evidence, they deduced that Jack Parsons must have accidentally dropped a coffee can full of the stuff. And that two-foot drop was all it took to end his life.

Rumors and stories continue to swirl around the memory of Parsons. Although he was not college educated, his innovations in rocketry during the 1930s and 1940s were amazing. His contribution in the field of rocket fuels was particularly important. In fact, a fair amount of credit is due Parsons for powering the nascent American space exploration program. Parsons is credited with inventing the process for casting solid fuel rocket motors.

Solid fuel rocket motors propelled the gigantic Saturn V rocket that carried American astronauts to the moon. Today, Parson's contribution is the basis for the design of the two solid rocket boosters that provide the liftoff thrust for every NASA space shuttle.

Parsons's personal life was as noteworthy as his professional career. He was intensely interested in mysticism and was rumored to be a disciple of Aleister Crowley, a British writer and perhaps the preeminent occultist of the 20th century. Before each rocket launch, colleagues noted that Parsons would recite Crowley's "Hymn to Pan," a strange bit of poetry dedicated to the Roman pipe-playing, horned god of fertility.

But Parsons's reputation as a risk-loving disinhibitionist went even further. Stories of orgies, black magic, even incest swirled around Parsons. If Parsons did worship Satan, then his choice of career suited him well, for there was likely no man alive more comfortable working with fire and brimstone. His work was not forgotten. In 1972 a crater on the dark side of the moon was named after him. Given his nature, that's probably a place he'd find desirable.

Type T Personalities

Social scientists have labels for Parsons and others like him. They call them Type T personalities. T stands for "thrill seeker," a high-energy personality who craves excitement and stimulation. When a thrill seeker can't find it, he or she creates it. Thrill seeking is a term that encompasses a great deal of mental territory. Thrills can be physical, or they can be mental.

Thrills can be more than just fun, however. Done often and done well, thrill seeking, as we'll see, imbues those who attempt it with a number of important attributes such as self-reliance, situational control, and the ability to think and act rationally under extraordinary circumstances. The key is to understand the balancing act that must occur for thrill seeking to be both artful and beneficial. Learning the art of living dangerously, I firmly believe, is an important life skill.

Put another way, people choose to fall into one of two risk categories. One can eschew risks or seek out risks. And those who choose to seek out risks can do it poorly, with malevolence or failure, or they can do it well, with art and elegance and a high chance of attaining their goals. The question is how to insure the outcome is the latter.

Dr. Frank Farley of Temple University has written extensively on the positive and negative aspects of thrill seeking and the associated personality types. Type T personalities, says Farley, exist on a continuum. On one end are the Big-T people, those who go out of their way to flirt with danger. It's not hard to find examples. Consider the unfortunate Parsons, Ernest Hemingway, DNA researcher Sir Francis Crick, the legendary explorer Amelia Earhart, social reformer Mary Harris "Mother" Jones, gangster Clyde Barrow, and gonzo journalist Hunter S. Thompson.

Including the felonious Barrow and the revered Crick in the same paragraph is significant. Thrill seeking correlates with both delinquency and creativity. Some people, Crick and Earhart, for example, are admired for the results of their thrill-seeking instincts. Others, such as Barrow and Parsons, came to violent ends. It's the environment in which one operates that makes the difference. A centered, wealthy, and mentally mature thrill seeker can turn to fast cars, sports, and other socially acceptable outlets. But poor and irresponsible ones may turn to criminal or antisocial behavior for outlet. For them the phrase "everything good is either illegal, immoral, or addictive" is sadly too true.

Taking the thought a step further, besides those who risk physical life and limb, there are those who risk it all mentally or academically. Farley cites Margaret Mead, Albert Einstein, and Helen Keller as famous risk takers who were thrilled by the adrenaline rush of their original thinking.

There are many famous Big-T people. In fact, while Big-T test pilots, explorers, and political leaders made great impacts on their societies, there are those who are famous for no reason other than their willingness to take on risk. Evel Knievel is an easy, if hackneyed, example. Daredevils, motorcycle jumpers, and circus performers are famous and admired for no reason other than what they do appears dangerous. There is no particular benefit to society in the work they do. We as a group have little practical use for lion taming a la Siegfried and Roy, or jumping rows of cars on a Harley XR-750 while wearing a patriotically themed, star-spangled jumpsuit.

On the other end of the spectrum are the little-t people. They are people, Farley told me, "who cling to certainty and predictability, avoiding risks and the unfamiliar. Such people are usually neither criminal nor creative; they're gray compared to the bold red of the Type T personality."

It's difficult to come up with examples of well known little-t's. After all, no little-t makes headlines or gets a biopic on cable television. The rare well known ones are fictional characters like Felix Unger, Caspar Milquetoast, and Bud Frump. The timid, inertia-bound title character imagined by T. S. Eliot in his famous poem "The Love Song of J. Alfred Prufrock" may well be the poster child for a life shortchanged by little-t thinking and ennui.

Most people fall somewhere between Knievel and Prufrock; neither as hungry for novelty and excitement as the extreme Big-T nor as mind-numbingly fearful as the frumpish little-t. Imagine a team of psychologists testing the world's population and determining every individual's attitude toward risk. While a census of this sort has never been done, many surveys have been completed. These studies provide a fair indication of what the universal thrill-seeking capacities and needs of the general population really are.

It's no surprise that, like so much else in nature, thrill-seeking behavior in the real world is modeled by what statisticians call a normal curve, the bell-shaped graph that shows how most things tend to cluster around a mean and that large differences from that mean are relatively uncommon. The people who place noticeably distant from everyone else are the outliers. Those rara avises are not the primary concern of this book.

So then, who is? I'll use a bit (don't worry, just a small bit) of statistical data to answer that.

The Golden Third

The normal curve describes the distribution of humankind's proclivity to seek out and enjoy new sensations. As I mentioned in the opening chapter, Dr. Marvin Zuckerman at the University of Delaware has performed extensive research on the psychology of risk taking. He designed a series of questions that measure the test taker's propensity to take risks, to seek out new experiences, and to avoid boredom. The instrument, called the Sensation Seeking Scale, has gone through several iterations and modifications. It remains the best known and most used statistical instrument for studying human behavior as it applies to risk taking. Thousands of subjects have taken the test, providing a reliable and scientifically valid statistical database.

The graph of sensation-seeking behavior follows a normal distribution, which means it resembles the mountaintop of a stratovolcano. The vertical height of the curve at any point indicates the frequency of a particular score. Thus the peak indicates the mean or most common level of sensation seeking in human

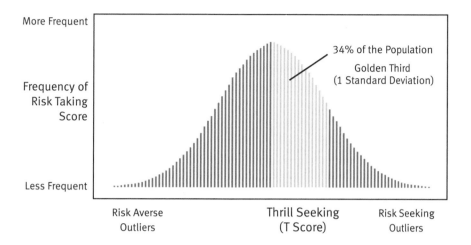

The normal curve of risk taking

beings. If a person's score falls to the left, then that person is more cautious, typically less willing to take risks, better able to handle routine, and less inclined to try new things.

In a normal distribution, such as this diagram depicts, people with scores placing them to the far left or far right are few compared with those clustered around the mean. The outliers, or those more than one standard deviation from the mean, show either a high propensity to seek new sensations or a strong desire to avoid the unfamiliar and the risky.

Standard deviation is a statistician's term that indicates a certain measure of nonconformity with those near the mean. As the diagram shows, the proportion of the population whose risk-taking behavior falls between the mean and the first standard deviation line is about 34 percent. Therefore roughly two-thirds of all people have sensation-seeking scores that fall between the mean and one standard deviation to the left or right.

This book is written for the people within one standard deviation of the thrill seeking mean, the 68 percent of the world's population that neither takes ridiculous chances nor is timorously risk averse. If you fall inside the Golden Third (the shaded section in the diagram between the mean and the first standard deviation, which will be explored in a subsequent chapter), then you're in the 34.1 percent of the world's population that lies within one standard deviation *upside* of the mean.

John Parsons most likely never took a psychological profile. But based on accounts from those who knew him, he would almost certainly place well right of the mean. His proclivity to take risks—mental, physical, and financial—made him interesting and out of the ordinary. The over-the-top and sordid aspects of his personal life aside, his capacity to court danger with a certain degree of flair and style is admirable.

Many artful and risk-tolerant people have made enormous contributions to society and have not been blown to bits in an explosion of homemade chemicals. These are the people of the Golden Third—the people who understand the art of living dangerously.

These people understand the dangers and rewards of sensible risk taking. Golden Thirders have attained a balanced, comfortable approach to trying out new activities and ideas that involve reasonable but significant risk. I consider this the optimum range on the risk-taking/thrill-seeking continuum.

However, I believe the information and ideas presented here will interest most people, not just those already in the Golden Third. If your makeup places you on the asymptote-hugging left side of the curve, you might not be able to internalize the arguments I present for consciously introducing more risk into your life. And if your makeup places you on the extreme right, well, you don't need my advice, other than to be careful out there. But if you reside anywhere near the middle of the risk-taking curve, as the vast majority of people do, this book is for you.

Eminent scholars in institutions worldwide have researched and written about the advantages that accrue to those who make an effort to move toward the right on the little-t/Big-T risk-taking continuum, at least until you recognize that you've reached your optimum risk-taking zone.

It is a sad fact that for many reasonable people, societal strictures have forced the information their intellect desires underground, unavailable in the mainstream media and untaught in schools and universities. I hope the information provided here may change things a bit for the better. Here, for the first time, the know-how and wherewithal go mainstream and into the minds and hands of you who want it.

In a nutshell, this book explores the theory and practice of reasonable risk taking. It is written to be useful for those on both sides of the risk-taking curve. If you find yourself already on the right side of the mean, the pages that follow provide information and projects that will interest, stimulate, elucidate, and educate.

If you operate on the left side of the statistical mean, then you will find advice and ideas for moving to the right that take (or keep) you in the region of statistically proven higher-than-average life satisfaction.

Such experiences are described by a single word: edgeworking.

2

What Is Edgework?

"*It was dangerous lunacy, but it was also the kind of thing a real connoisseur of edgework could make an argument for.*"

—HUNTER S. THOMPSON,
FROM *FEAR AND LOATHING
IN LAS VEGAS*

I've written this book for readers who want more out of life: a bit more brain stimulation, a little more excitement, perhaps a few more things to talk about at the next cocktail party or social gathering. In the next section, I'll make my case as to why reasonable, eyes-wide-open, and carefully considered risk taking is good for you. My hope is that this book opens your mind to the world of interesting, edgy activities that require no great investment in time or money.

Hunter S. Thompson, perhaps the most extreme Big-T writer in recent memory, termed these sorts of activities "edgework," and he wrote extensively on his pursuit of such experiences. Thompson explored the areas where only a thin, knife-edged boundary exists between legal and illegal, moral and immoral, life and death, and funny and just plain stupid. In my opinion he did this more vividly and more transcendently than any other writer.

But, with due deference to the late, great gonzo journalist, living life in the Golden Third precludes incorporating the Thompson hallmarks—shotguns, LSD, and anarchy—into the lifestyle. It's hard to make them artful. And there is a better path: the art of living dangerously. Living dangerously is an art and a science. Since artfully dangerous and non-artful edgework activities share some similar characteristics, it's tempting to simply define the art of living dangerously by a loose "I know it when I see it" definition. However, to be helpful, the concept needs to be reined in, and some structure placed upon it. Upon a great deal of reflection, artful, dangerous living is not all-encompassing. It can be understood based on the extent to which it possesses some key characteristics. Artfully dangerous activities:

- have short learning curves;
- are human focused, as opposed to technology centric;
- are not unduly expensive; and
- demonstrate true but reasonable risk.

Selecting these criteria was difficult. Arguments certainly can be made for including other criteria or for excluding one or another of these. But to me, the difference between artful edgework and less desirable, dodgy activities is that edgeworkers are not burdened with a long learning curve, a great deal of complexity, or the need to spend a lot of money. Let's explore each of the qualities that make up quality edgeworking.

Short Learning Curve

An artfully dangerous activity does not require long-term or expensive training before the practitioner can even attempt it, nor should it require the level of planning associated with organizing a Himalayan trek. In fact, the artful component of living dangerously frequently requires a spirit of playfulness and a conscious effort to eschew seriousness and self-importance.

Human Focused

In *Devices of the Soul,* author Steven Talbott argues that real experiences are always preferable to virtual ones. "Studying Antarctica on the Internet," he recently stated, "pales in comparison to taking a magnifying glass and studying bugs in your backyard." Digital technology, he concludes, delivers a weak and inferior replacement to actual hands-on experience.

I believe this. It's not just computer use that diminishes human artfulness; it's overreliance on machines of any sort. I find that the most enjoyable activities are not heavily mechanized. To me this means that an artfully dangerous activity does not involve a complex machine or technological addition to the human body, such as a parachute or an aqualung.

Inexpensive

This tenet is a corollary to being human focused. Artfully dangerous activities generally don't need to cost a lot of money. While there is nothing wrong with spending money on interesting things, say, transatlantic balloon trips or space tourism, it just doesn't seem artful. Further, such priciness almost inevitably changes the activity from human focused to technology focused.

Involves Reasonable Levels of Risk

While dangerous and artful activities have a real and noticeable germ of danger at their core, they are far more reasonable in terms of safety than true outlier practices, which may be ill-conceived or extremely dangerous—or both. In the Golden Third, risk is present, yet moderate. Part of the appeal of living dangerously may be that there is a real possibility of death. However, that possibility should be extremely, extremely remote. It's the difference between driving fast and focused on an autobahn and driving drunk on a freeway; the difference between savoring chef-prepared tiger puffer fish and eating lukewarm, raw chicken.

Some readers may notice that I haven't included extreme sports such as skydiving and scuba diving in this book. This is not because they are too risky. My exclusion of extreme sports is due to the amount of training and equipment they require. The same equipment that lets devotees slip the fetters of gravity or oxygen simultaneously shackles them to long periods of training and inevitable expense. And, having obtained the requisite skills and equipment, the user must operate within the bounds of approved practice.

Mountain climbers, cave divers, and BASE jumpers are undoubtedly Big-T individuals. But their choice of activity involves being exposed to the elements, thereby requiring the use of specialized equipment to control an inherently uncontrollable and unpredictable environment. Interesting? Yes, but it isn't the area of risk taking covered here. In *Waiting for the Weekend*, author Witold Rybczynski observes that sporting activities, taken to an extreme, become laden with a sense of seriousness that counters the original, playful intention. Such seriousness changes the endeavor from play to work. Enjoyable work perhaps, but work just the same.

Instead, a high proportion of the projects and experiences found here involve high velocity, but low costs to entry and even lower technology. Much of this book explores the intersection of risk taking and creating simple yet interesting things.

3

Where the Action Is

*"I don't want to come to the end of my life and find that
I have just lived the length of it. I want to have lived the
width of it as well."*

—Diane Ackerman

Where do you reside on the risk-taking scale? This is a surprisingly difficult question to answer. We all have preconceived notions of our proclivity toward taking risks. Much of the
time, we're wrong.

"I'm a rootin'-tootin', hard-living bad boy," one fellow may
say, pointing to his battle scars and tattooed torso as proof. That
may be, but studies show that it is difficult for most people to
make an accurate self-assessment without help. Indeed, you may
view yourself as a James Bond—like suicide commando, while your
friend sees herself as a safety-obsessed health fanatic. And
chances are, you both may assess your appetites for risk incorrectly compared with the rest of the world.

Dr. Zuckerman's book *Behavioral Expressions and Biosocial Bases of
Sensation Seeking* provides some direction toward accurate self-
assessment. Over decades, Zuckerman developed a series of psy-

chological test instruments called the Sensation Seeking Scales. They consist of a series of activity-related questions. Comparing their results to a very large database, test takers can quantitatively determine their penchant for risky living.

I've adapted Zuckerman's test with the intent of making it fast and easy to take. I don't claim it to be a perfectly accurate window into your inner psyche; in fact, this is nothing more than a simple and quick starting point to evaluate your risk-taking proclivity. By providing accurate and truthful answers to the questions, you can assess your place on the living dangerously normal curve vis-à-vis a large sample of other test takers. That's an important thing to know about yourself.

The test below is based on portions of Dr. Zuckerman's Sensation Seeking Scale Form VI (SSS-VI) questionnaire. But unlike more rigorous tests used in academia and social psychology circles, this quiz is intended to measure just one aspect of a person's total risk-taking behavior, a single dimension known as "thrill and sensation seeking." Don't attempt to read too much into this test; psychologists typically measure a number of additional factors when assessing a person's overall attitude toward risk taking.

This test will provide a baseline on which to judge your current attitudes toward living dangerously. Read the instructions and answer all of the questions before turning to the scoring instructions.

The Thrill and Experience Seeking Self-Evaluation

Part I: Experience

Below is a list of many different activities. Please indicate whether you have actually engaged in each activity. Answer all items with one of the following:

A. I have never done this.

B. I have done this once.

C. I have done this more than once.

Please mark only one response for each of the items: A, B, or C.

Answer all of the items. There are no wrong or right answers. The only interest is in your experiences, not in how others might regard these activities.

Climbing steep mountains	(A)	(B)	(C)
Running in a marathon	(A)	(B)	(C)
Walking a tightrope	(A)	(B)	(C)
Parachute jumping	(A)	(B)	(C)
Flying an airplane	(A)	(B)	(C)
Scuba diving	(A)	(B)	(C)
Horseback riding at a gallop	(A)	(B)	(C)
Sailing long distances	(A)	(B)	(C)
Swimming alone far out from shore	(A)	(B)	(C)
Skiing down high mountain slopes	(A)	(B)	(C)
Exploring caves	(A)	(B)	(C)
Racing cars	(A)	(B)	(C)
Backpacking in Europe	(A)	(B)	(C)
Snorkeling over a reef	(A)	(B)	(C)
Backpacking in the wilderness (in the United States)	(A)	(B)	(C)

Part II: Intentions for the Future

Be sure you read the new instructions for this part.

Below is a list of many activities. Please indicate your degree of interest on engaging in each activity in the future, regardless of whether or not you have engaged in the activity in the past. Answer all items using one of the following options: In the future,

A. I have no desire to do this.

B. I have thought about doing this but probably will not do it.

C. I have thought of doing this and will do it if I have the chance.

Please mark only one response for each of the items: A, B, or C. Be frank. Answer all of the items.

Climbing steep mountains	(A)	(B)	(C)
Running in a marathon	(A)	(B)	(C)
Walking a tightrope	(A)	(B)	(C)
Swimming the English Channel	(A)	(B)	(C)
Parachute jumping	(A)	(B)	(C)
Flying an airplane	(A)	(B)	(C)
Scuba diving	(A)	(B)	(C)
Horseback riding at a gallop	(A)	(B)	(C)
Sailing long distances	(A)	(B)	(C)
Swimming alone far out from shore	(A)	(B)	(C)
Climbing Mount Everest	(A)	(B)	(C)
Skiing down high mountain slopes	(A)	(B)	(C)
Exploring caves	(A)	(B)	(C)
Hunting lions or tigers	(A)	(B)	(C)
Racing cars	(A)	(B)	(C)
Backpacking in Europe	(A)	(B)	(C)
Traveling in Antarctica	(A)	(B)	(C)
Taking a trip to the moon	(A)	(B)	(C)
Snorkeling over a reef	(A)	(B)	(C)
Backpacking in the wilderness (in the United States)	(A)	(B)	(C)
Traveling up the Amazon	(A)	(B)	(C)
Surviving alone on an island for a week	(A)	(B)	(C)

Interpreting Your Results

For each part, the responses are weighted as follows:
A = 1 point; B = 2 points; C = 3 points.

Using the weights above, add the number of points. Then use the graphs below to find your rank in relation to Zuckerman's sample population. For example, let's say you answered A to the first seven questions in Part I and B to the second eight questions. Your Part I total would be $(7 \times 1) + (8 \times 2) = 23$.

Note that there are different graphs for men and women. Note also that you will end up with two scores, one for experience and one for intention. The test is intended to measure the full breadth and variety of your experiences and thrill-seeking desires.

To interpret your two scores, locate your point total on the x-axis of the relevant experience and intention graphs. The farther right you place on the graphs, the greater your thrill and sensation seeking needs and the bigger Type T personality you possess.

For example, if you're a male and your scores are 16 and 35 for the experience and intention questions, respectively, you place a bit below the average of risk-taking proclivity; that is, you tend towards the little-t side of the spectrum. If you score 16 on experience and 45 on intention, you're more inclined to take risks than your life history would indicate to this point.

If you score, say, a 36 on the experience section and a 65 on the intention section then, my friend, you're among the biggest of the Big-T's (and you'll probably have a hard time finding a life insurance policy).

If you are female and your scores are 22 and 42, respectively, then lucky you, you fall within the Golden Third, where the action is.

Dr. Zuckerman cautions SSS-VI test takers to be aware that most people have not experienced a majority of items on the experience inventory. Therefore the distribution may be skewed toward the low end, and an individual's scores, particularly those based on experience instead of intention, may skew quite low. Thus, you may be a bigger risk taker, edgeworker, and Big-T personality than your experience test score indicates.

Test interpretation graphs

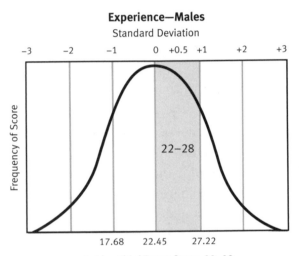

Experience—Males
Standard Deviation

22–28

Golden Third Target Score: 22–28

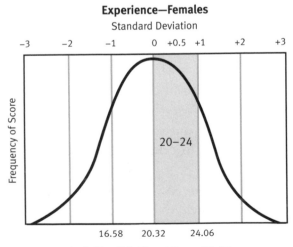

Experience—Females
Standard Deviation

20–24

Golden Third Target Score: 20–24

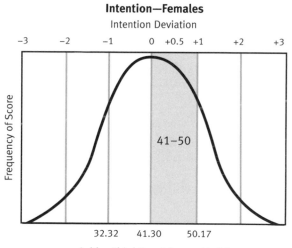

Intention—Males

Intention Deviation

Golden Third Target Score: 45–55

Intention—Females

Intention Deviation

Golden Third Target Score: 41–50

Once you take the test and determine your position on the sensation-seeking scales, what value does that information provide? That's the subject of the next chapter.

4

Why Live Dangerously?

In the previous chapter, I presented a method for gauging your inclination for living dangerously and introduced the idea that, to a point, the ability to wage risks is a useful and worthwhile attribute. In this chapter, we explore benefits that accrue from a lifestyle that includes absinthe and flamethrowers, and why living dangerously and artfully is an idea worth some reflection.

First, artful yet dangerous living builds your repertoire of life experiences and your store of practical knowledge. Almost every new, varied, and unusual experience you have makes you more valuable on some level because it builds within you something called "deep smarts."

What are deep smarts? Dorothy Leonard, professor emeritus at the Harvard Business School, and Walter Swap, professor of psychology emeritus at Tufts University, coined the phrase in their well-regarded book by the same name. They define deep smarts as the accumulated know-how and intuition gained through extensive and varied experience. It's expertise in the form of practical wisdom. Leonard and Swap tell us that it's just such deep smarts that provide the foundation for an individual's overall success.

It appears from their research that nurturing your risk-taking skills should result in personal experiences stored deep in your intellect, which have potential for personal value greater than any book learning. According to Dr. Swap, "Someone who may have a lot of book learning but not a lot of real life experience won't be able to look at a new situation and say, 'Aha, that reminds me of a time when I did X,' and that memory suggests a way to act. A person with deep smarts may not actually be able to recognize or to put into so many words where that knowledge came from—but is nonetheless able to react quickly and wisely."

Temple University's Frank Farley takes the idea even further when he declares that risk taking is one of the most important factors that positively influence mental growth. Risk taking requires self-confidence, he says, and that leads to the realization that you can do things you previously didn't think possible.

Taking risks proves your mettle; it displays your inner bravery. And successful outcomes in risky propositions lead to a well-deserved sense of accomplishment. These aren't just my opinions. There is a considerable body of scientific evidence backing up this assertion. At least two rigorously conducted scientific studies statistically show that, in the long run, people who take reasonable risks are happier and more successful.

In their study entitled "Characteristics of Risk Taking Executives," Kenneth MacCrimmon and Donald Wehrung of the University of British Columbia assert that "*a higher degree of success (i.e. being wealthier, obtaining a higher position, etc.) differentiated the risk takers from the risk averters* [italics added]. We hypothesize that for most businesses a person gets to the top by taking risks and having them work out for the best. The person who does not take risks is unlikely to get to the top."

In their popular book *Taking Risks,* MacCrimmon and Wehrung went further and laid out this conclusion in unambiguous terms: "We examined the relationships between risk and a variety of personal, financial, and business characteristics. Here is what we found: *The most successful managers took the most risks.* The more rigorously success was defined, the stronger the relationship became."

Evidence supporting this conclusion is widely distributed. In 2005 the German Institute for Economic Research conducted a highly publicized study in Germany. Entitled "Individual Risk Attitudes: New Evidence from a Large, Representative, Experimentally Validated Survey," the study supports the conclusion reached by the Canadian researchers. Dr. Armin Falk and his colleagues analyzed the risk-taking behaviors of a scientifically selected sample of the German population that closely approximates the characteristics of the country as a whole. In this huge, expensive, and complicated undertaking, a highly refined and engineered set of survey questions was administered to more than 22,000 carefully selected individuals. The purpose was to better understand the correlation between a person's risk-taking proclivity and a host of different personal attributes and inclinations.

The scientists were able to show how an individual's risk-taking attitude—basically their location on the little-t/Big-T continuum—predicted a large number of other attributes. Some relationships were not surprising. For example, Big-T types collected far more traffic tickets than little- t's. But other findings did surprise. Perhaps no conclusion stood out as much as the one relating risk-taking proclivity to general life satisfaction. Wrote the authors, "We find a strong positive association between life satisfaction and willingness to take risks in general."

Learning to Live Dangerously and Artfully

People can learn to increase their artful risk-taking abilities, says Temple University's Frank Farley. Those who do so gain self-confidence and are imbued with the belief that they control their own destinies. A number of academic researchers, including Professor Stephen Lyng of Carthage College, have come to a similar conclusion. Lyng edited an important academic work entitled *Edgework: The Sociology of Risk Taking*.

Lyng, a one-time motorcycle racer and skydiver, conceptualizes edgework as a form of boundary negotiation; a purposeful, personal expedition to explore the dividing lines or boundaries between that which an individual finds scintillating and that which is horrifying.

Lyng says that edgeworkers are highly aware of their personal capabilities and resources. Living dangerously, he told me, develops a sense of self-reliance in individuals. Such people understandably take great pride in their character, that is, in their ability to maintain command over chaotic situations and in conditions other people would regard as entirely uncontrollable. Developing this type of competence centers on the ability to avoid being paralyzed by fear, irrespective of the circumstances. It requires learning how to focus your attention on what is required to survive the situation. In other words, says Lyng, people who live dangerously become mentally tough.

According to both Lyng and Farley, people can consciously become more comfortable with risk. Not like Clark Kent, a quiet, assuming, introverted newspaper man who would step into a phone booth and transform from a little-t to a Big-T in an instant. But you can move along the continuum toward a Bigger T.

"As you try out things, you'll get to your risk tolerance," Farley continues. "My advice is start in the shallow end of the pool and work towards the deep end until you find your threshold."

Clearly the ability to take risks has its good side. But taken to extremes, it can also be imprudent, even stupid. Individuals who engage in too much and too extreme risky behavior are a danger to themselves, and worse, may possibly be a danger to others. It's true that the biggest Big-T types may do wonderful things, but it's just as true that crossing the line into outlier territory can be personally disastrous. Sure, we may admire Amelia Earhart, Steve Fossett, and even Jack Parsons, but they lived too fast and died too young. When they consciously chose their places as risk-taking outliers, their early demise was almost a foregone conclusion.

On the other hand, you might be a naturally cautious person, in the 50 percent of the population to the left of the mean in terms of risk-taking behavior. As the evidence shows, you'll do well to move to the Golden Third. The projects and practices described on the following pages are a fine way to get started.

If you do make a conscious effort to choose your risk-taking level, the place of choice is within the exciting yet reasonable confines of the Golden Third. This wonderful, sensible, and exciting area under the curve is where risk and reasonably long life expectancy coexist in rational proportion.

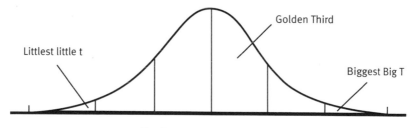

Risk Taking Propensity

In the pages that follow, I hope to provide information on the artfully dangerous path that eschews the outlier positions and instead promotes activities falling within the boundaries of the Golden Third.

For those of somewhat meek and timorous habits, I offer this information, hopefully to expand your experiences and knowledge, thereby providing you with the advantages of deep smarts. Those already comfortable with taking reasonable and calculated risks will find projects that further broaden their store of already considerable knowledge. And you'll have a good time while adding to your deep smarts.

Where do you go from here?

Having determined your personal baseline for edgeworking, you can attempt endeavors that stimulate personal growth and self-confidence. Some of the do-it-yourself projects may seem quite safe, even tame, except for some nagging uncertainties. Others fairly drip with adrenaline.

In addition to the projects, suggestions, and practices, you'll note a number of tips and guidelines. These distillations of edgeworking knowledge are set apart from the rest of the text in boxes. These tips are especially valuable because I interviewed the world's leading experts in their particular field. Each is recognized by peers as a standout, and is thought to be a leader in his or her sometimes arcane but always interesting and risky niche.

Part II

How to Live Dangerously

5

The Most Important Chapter in the Book

Here comes the first of several warnings that my publishers and lawyers rightfully insist on, so pay attention. You must understand that many of the activities and projects in this book include an element of risk that simply cannot be designed, directed, or done away with completely.

I've designed the projects to be as well thought-out and engineered as I could, testing them and adjusting them until I felt satisfied. But no one is perfect. It is possible, even likely, that there are gaps or errors in these pages. If errors are identified after publication, relevant information will be placed on the Web site www.AbsintheAndFlamethrowers.com. But I cannot and do not claim that the projects that follow are 100 percent safe and correct.

Consider yourself informed: this book is but a guide and, ultimately, the responsibility for your actions is no one's but yours. If you do not agree, put down this book; it is not for you. There is no shame in that. If you do go forward with the projects here, it's essential to understand that you and you alone are responsible for your decision to live dangerously.

Scary stuff, no? Should you attempt any of these projects? I'd love to tell you to plunge right in, but truly, it depends on your own tolerance for risk and your ability to build and manipulate things. These projects aren't for everybody. Some in fact require a good deal of forethought. But even if you decide against taking the plunge, the read will do you good. You'll find the information a worthy addition to your collection of knowledge. Based on my experience as a magazine writer and author of five DIY books, reading about out-of-the-ordinary projects can be pleasurable, even if you have no particular intention or timeframe for carrying out the project. And those of you who live in the Golden Third—you with a bit of daredevil in you—should you attempt even a couple of the projects, you will find them challenging and engaging.

In addition to projects, several dangerous practices are described. Some of the practices may strike the casual reader as unusual, but each in its own way is educational and interesting. And more than that, each has a certain honor, flair, a *je ne sais quoi* that you'll truly understand only if you actually attempt it.

If you decide to participate in the dangerous practices described, you may find it helpful to find a knowledgeable practitioner to guide you through the difficult parts.

Safety

If you're like me, you'll likely jump from page to page, from project to project instead of reading the book from start to finish. If that's your style, great. Just be sure you read this section completely.

Your safety is paramount. I've spent a lot of time developing projects, and while I can't guarantee you'll never get hurt, I believe you're better off to build upon what I've already found out, sometimes the hard way.

These are the universal guidelines. Each chapter also contains additional recommendations specific to each project or practice.

Wear safety goggles, safety glasses, or face shields as appropriate. Designed so they won't shatter or crack, the lenses of safety eyeglasses are much tougher than your regular glasses.

If chemicals are involved, always work in a well-ventilated area.

Observe good housekeeping practices. Keep your work area clean and tidy: it's safer—cutting down on burns, trips, spills, and other accidents—and you'll spend less time looking for things. Wash your hands with soap and water after handling chemicals. Store fuels and oxidizers in separate, secure cabinets. Dispose of all chemical waste properly.

Never grind *mixtures* of chemicals in a coffee grinder or mortar and pestle. Always mill chemicals individually.

Read labels and equipment instructions carefully before use.

Keep dangerous materials and tools away from children. Store dangerous chemicals and tools in an appropriate, secure locker.

Know the locations and operating procedures of all safety equipment. This includes your first aid kit and fire extinguisher.

Always handle flammable or dangerous materials with great care. Keep them away from all flames and potential ignition sources. If something gets in your eye or on your skin, immediately flush with water for at least 20 minutes. Don't touch chemicals with your fingers. Do not eat any chemicals no matter how tasty they smell.

Be certain all safety guards are in place before operating any machine or equipment.

Dress properly for the activity. Don't wear loose or baggy clothing around machinery. Shoes must completely cover your feet. Always wear long sleeves when welding or grinding, because sparks or chemical ejecta are possible. Wear a lab coat or apron when your task involves chemicals. Wear a hat when practicing with a bullwhip.

Don't assume that substitutions, variations, or other changes from the directions provided will be safe or provide the expected result. So, before you write to me to ask, "Can I use match heads for a rocket propellant? Can I skip wearing safety glasses? Can I prepare my own pufferfish?" I'll tell you that the answers to those questions are no, no, and God NO!

The information in this chapter and in this book is not all-encompassing. Look around your shop before experimenting or building, and take stock of the situation. Then do what you deem necessary to work safely.

Dress for success. Here is a summary of the safety gear I recommend you wear when working with chemicals and tools. Here's a tip: buy the most stylish safety equipment you can find. If your gear has style and makes you feel cool, you'll be more inclined to consistently wear it.

Take a moment to read and understand this chart:

SAFETY EQUIPMENT

When You're Doing This:	Watch Out for:	How to Protect Yourself:
Working with Metal: Chipping, grinding, hammering, machining, sanding	Flying metal chips and fragments Particles Sparks	Safety glasses or goggles Face shield* Lab coat or apron Leather gloves (if they don't interfere with the operation)
Working with chemicals	Splashes Fumes Vapors	Safety glasses or goggles Face shield* Lab coat or apron Dust mask or respirator Rubber gloves
Tamping propellants	Explosions Liquid splashes Fumes	Leather Gloves Dust mask or respirator Safety glasses or goggles Face shield* Lab coat or apron Hair covering Explosion bunker
Heating chemicals Working with explosives Working with wood	Dust	Goggles Dust mask

* Always wear a face shield in combination with glasses or goggles.

Prior to starting any project, check the companion Web site at www.AbsintheAndFlamethrowers.com for important updates and safety information.

6

Obtainium

What the heck is Obtainium? It may sound like a cross between osmium, barium, and titanium, but it's not. It's the stuff you need to obtain—parts, chemicals, and equipment—to participate in the projects and practices that follow. This chapter provides basic information on finding what you need. First, some background.

Several years ago, in the course of researching *Adventures from the Technology Underground*, I interviewed a number of machine artists in San Francisco. I enjoyed every minute I spent with them. Edgy, creative, dexterous, and artistic, these folks are members of a loosely defined artistic community whose specialty is transforming machinery into works of artistic expression. Their exhibitions include monstrous machines and seemingly dangerous performance installations, often involving fire-breathing robots, unguarded spinning blades, and other decidedly non–OSHA-compliant arrangements of industrial equipment—typically accompanied by an over-amplified, four-to-the-floor drum machine beat. Endemic in this community is a spirit of anarchy, evolving as it did from Dadaism and original, unconstrained artists such as Marcel Duchamp and Jean Tinguely.

One thing I wondered is how contemporary machine artists obtain parts and raw materials for the sculptures and figures. Many artists are tight-lipped about their techniques for procuring such arcana as high-torque motors, large capacitors, massive quantities of conductive wire, pneumatic cylinders, and various drivetrain parts. The market for machine art being what it is, I couldn't fathom how the artists could afford the requisite raw materials. But necessity, as well as creativity, has driven the machine art community to find affordable sources. As experts in understanding surplus equipment markets, they have become adept at delving into the guts of big machines to get to the useful and valuable component parts. On rare occasions they turn to other methods.

I met one rough looking character called "Catman" by those who knew him. Catman told me how, prior to his successful rehab, he roamed the junkyards and construction sites of the Bay Area at night with a shopping cart borrowed from the A&P. He took wire and promising looking metal parts, which he called Obtainium, in the hope of selling the parts to make enough money to buy fortified wine.

But this, I think, was atypical. For the majority of their creations, most of the artists I met use more ethical and reasonable methods to obtain supplies of a chemical, mechanical, or electrical nature.

While larceny is unacceptable, the spirit is worth keeping. This chapter covers the acquisition of elemental Obtainium, the basic ingredient for dangerous-living projects. Obtainium comes from many places. As described in the paragraphs that follow, take heed and attend to your safety, for every use of Obtainium is "off-label" and requires thought and consideration before incorporation.

Chemicals and Laboratory Supplies

Obtaining chemicals and laboratory equipment is becoming more difficult. Books of chemistry experiments from the 1940s and 1950s often advise visiting the local drugstore to obtain chemicals such as sulfur, saltpeter, and sulfuric acid. Those days are gone, I fear. On the few occasions when I asked for such items at the counter, my request was greeted with blank looks or shrugs.

While few retail stores sell supplies such as chemical oxidizers, electronic scales, and cannon fuse, a number of mail-order companies do. However, government authorities, claiming a need for ever-increasing national security, continually seek to restrict the law-abiding citizen's access to what would seem to be perfectly reasonable materials.

At the time this book was written, all of the chemicals described are relatively easy to obtain. However, there are reasonable laws in place regulating the quantities, shipping methods, and availability of certain chemicals. Legitimate retailers will always explain any relevant restrictions before you buy. Listed here are several chemical and laboratory equipment suppliers.

- Skylighter, PO Box 480, Round Hill, VA 20142, www.skylighter.com. Sells supplies and chemicals for pyrotechnic hobbyists. Stock includes potassium nitrate, sulfur, and charcoal in reasonable quantities.
- Firefox Enterprises, PO Box 5366, Pocatello, ID 83202, www.firefox-fx.com. Similar merchandise to Skylighter.
- United Nuclear, PO Box 851, Sandia Park, NM 87047, www.unitednuclear.com. Smaller retailer of unusual scientific items. Carries a variety of difficult to find chemicals that others often don't stock.

- CannonFuse, PO Box 222, Milltown, MT 59851, www.cannonfuse.com. Good source of Visco and other fuses.
- Cannon-Mania, PO Box 552, Stratford, CT 06615, www.cannon-mania.com. Another good source of cannon fuses.
- Ward Scientific, PO Box 92912, Rochester, NY 14692, www.wardsci.com. Large, all-purpose scientific supply house that sells chemicals, lab equipment, and safety equipment. Large inventory and selection.
- Sargent Welch, PO Box 4130, Buffalo, NY 14217, www.sargentwelch.com. Similar to Ward Scientific.
- Amazon, www.amazon.com. I've found that many of the chemicals and other supplies specified in this book can be purchased through Amazon's Web site. Amazon has thousands of affiliated vendors that sell everything from potassium nitrate to sorbitol.
- Check www.AbsintheAndFlamethrowers.com for updates on where to procure materials and tools.

Material Safety Data Sheets

When you buy chemicals, it is important that you read the relevant Material Safety Data Sheet (MSDS) to learn how to handle them safely. The MSDS is a standardized fact sheet in which manufacturers describe the chemical properties of a product. Typical information includes physical data (solid, liquid, color, melting point, flash point, etc.), health effects, first aid, reactivity, disposal, spill and leak procedures, and personal protection. The U.S. Occupational Safety and Health Administration requires that this information be provided to workers who may be exposed to chemicals while on the job. The information is also essential for experimenters.

The MSDS is important because it describes the hazards of a material and provides information on how to safely handle, use, and store the material. Since the MSDS cannot contain all the information on any chemical, you also must read and follow package labels carefully.

Spectrum Group Division of United Industries P. O. Box 142642 St. Louis, MO 63114-0642 **Material Safety Data Sheet** Complies with OSHA's Hazard Communication Standard, 29 CFR 1910.1200	**Hazardous Material Identification System – (HMIS)**	
	HEALTH – 1	REACTIVITY – 0
	FLAMMABILITY – 0	PERSONAL – Rubber gloves, boots

I Trade Name: Spectracide® Stump Remover

Product Type: Granular Stump Decomposer

Product Item Number: 56420	**Formula Code Number:**

EPA Registration Number	**Manufacturer**	**Emergency Telephone Numbers**
N/A	Chemsico Division of United Industries Corporation 8494 Chapin Industrial Drive St. Louis, MO 63114	For Chemical Emergency: 1-800-633-2873 For Information: 1-800-917-5438 Prepared by: C. A. Duckworth Date Prepared: November 12, 2001

II Hazards Ingredient/Identity Information	**III Physical and Chemical Characteristics**
Chemical % OSHA PEL ACGIH TLV Potassium Nitrate 100.0 NE NE CAS #7757-79-1	Appearance & Odor: White to off-white granules. No odor. Boiling Point: NA Melting Point: 633ºF Vapor Pressure: None Bulk Density: 78 lb/ft³ Vapor Density: NA % Volatile (by vol.): NA Solubility in Water: 31g in 100g water

IV Fire and Explosive Hazards Data	**V Reactivity Data**
Flash Point: NA Flame Extension: NA Flammable Limits: NA Autoignition Temperature: NA Fire Extinguishing Media: Water fog Decomposition Temperature: 752ºF Special Fire-Fighting Procedures: Do not use Dry Chemicals, Carbon Dioxide or Halogenated agents. Unusual Fire & Explosion Hazards: This material is an oxidizer which may support burning or explosion when mixed with combustible materials and ignited.	Stability: Stable under normal storage conditions. Polymerization: Will not occur Conditions to Avoid: May burn vigorously or explode when mixed with combustible materials and ignited Incompatible Materials: Hazardous Decomposition or Byproducts: NA

VI Health Hazard Data	**VII Precautions for Safe Handling and Use**
Ingestion: Harmful if swallowed. First Aid: Give one or two glasses of water or milk and call a physician. Eye Contact: May cause irritation. First Aid: Rinse with plenty of water. Call a physician if irritation persists. Skin Contact: May cause irritation. First Aid: Rinse with plenty of water. Call a physician if irritation persists Health conditions Aggravated by Exposure: None under normal use. Ingredients listed by NTP, OSHA, or IARC as Carcinogens or Potential Carcinogens: None	Steps to be Taken in Case Material is Released or Spilled: Avoid contact with granules. Sweep up and place in waste container. Avoid contact with combustible materials. Waste Disposal: Do not reuse container. Place container in trash. Handling & Storage Precautions: Store in a cool area.

VIII Control Measures	**IX Transportation Data**
Read and follow label directions. They are your best guide to using this product effectively, and give necessary safety precautions to protect your health.	DOT Shipping Name: Stump decomposer DOT Hazard Class: None

The information and statements herein are believed to be reliable but are not to be construed as warranty or representation for which we assume legal responsibility. Users should undertake sufficient verification and testing to determine the suitability for their own particular purpose of any information or products referred to herein. NO WARRANTY OF FITNESS FOR A PARTICULAR PURPOSE IS MADE.

Material Safety Data Sheet example

The Household Products Database

Obtainium may be as close as under your kitchen sink or on a shelf in your garage. But how do you know what's in the bottles and jars? A good resource is the Household Products Database. Hosted on the federal government's National Institutes of Health Web site, this database contains health information on thousands of chemicals used by consumers—everything from adhesives to weed killer. The information is gleaned from many publicly available sources including the labels on bottles and manufacturers' MSDS.

In addition to safety information, the database contains quantitative information on the chemicals in the product. This makes it possible to search by chemical name to find a consumer product that contains the particular ingredient you're seeking. For example, searching the HPD for potassium nitrate turns up a couple of consumer products formulated from 100 percent potassium nitrate. While buying stump remover may be no less expensive than ordering potassium nitrate by mail, the former may be more convenient and faster.

Depending on the project, the purity of chemicals and substances can be of great importance. In some applications, fireworks for example, extreme purity is of less concern. How do you know what you're buying when purity is critical?

Guide to Purchasing Chemicals

Manufacturers use standard descriptions to denote the quality of their products. Here are the different chemical classifications in order of purity, from highest to lowest.

American Chemical Society (ACS) Grade:	This is the purest, highest quality grade of chemical normally available. It's also the most expensive, and it meets the exacting standards set by American Chemical Society.
Reagent Grade:	The purity of reagent grade is generally equivalent to ACS. Chemists often use reagent grade chemicals for analytical and quantitative work. It's very good for general lab use and also expensive.
Laboratory Grade:	This is of intermediate quality. The exact amount of impurities may not be known. However, this grade is usually pure enough for most amateur laboratories.
Pure Grade:	Sometimes referred to as purified or practical grade, this is a lower-level intermediate quality. Although this grade does contain impurities, it is pure enough to meet the requirements of most amateur projects.
Technical Grade:	Used industrially, this somewhat lower-quality grade may have higher levels of impurities than reagent or laboratory grade chemicals.

Choose the most appropriate grade based on need and cost. I typically choose laboratory grade chemicals as my default level of quality, although technical grades often work satisfactorily.

How to Talk with Parts Suppliers

Some parts distribution companies are easier to work with than others. Good ones welcome calls from individuals and happily accept small orders. Not-so-good ones don't think you're worth the bother. You can find out quickly which side a company is on by determining its minimum order size and/or minimum shipping charge.

In my experience, companies with online catalogs are the easiest to work with. Information on pricing and product availability is accessible instantly. You can open multiple browser windows to compare prices among several catalogs at once.

Quite often you'll need to call the company with questions. In that case, make sure you're well prepared. Prior to making the call, have a clear idea of the information you need and have your credit card number handy. That way nobody's time gets wasted.

You may or may not want to provide information about your particular project. I typically simply say, "My name is William Gurstelle. I'm an engineer, and I'm working on a prototype project. Can you help me order some parts?" Most of the time that's all it takes. If not, just call someone else.

Where to Buy Parts

The companies listed here represent what I consider to be the cream of the crop in terms of being easy to do business with. I have several criteria. Reputation is critical. I've used the organizations below frequently and have rarely had a problem. On those rare occasions when problems did arise, the companies took care of them.

Several of these organizations ship orders the day they receive them, and most feature well-organized Web sites with online catalogs that make ordering a snap.

Mechanical Parts and Materials

McMaster-Carr
My personal favorite because of the huge number of items it carries and the ease with which I can do business with the company. McMaster-Carr ships quickly, offers online ordering, maintains a

two-year history of orders, and requires no return material
authorization or restocking charge on returned parts.

Enco
Another good choice for tools, mechanical parts, and other general shop supplies. Enco claims its private label products are
often lower priced than many competitors' similar products. The
company ships quickly, has a reasonable return policy, and
answers its customer service phones on Saturday.

Small Parts
Not surprisingly, Small Parts specializes in small quantities of
parts for making samples and prototypes. The company carries a
great number of specialty raw materials that may be hard to find
at the larger, more general parts suppliers.

Online Metals
Online Metals' selection includes aluminum, copper, brass,
stainless steel, bronze, hot and cold roll steel, titanium, and
industrial plastics. Since it cuts materials to length, you buy only
what you need. The company typically ships quickly.

Electrical Parts and Materials

Mouser Electronics
Mouser ships quickly, has nearly a million electronic parts in its
catalog, and is geared toward design engineers and hobbyists.

Radio Shack
Radio Shack stores stock a small selection of commonly used elec-
tronic parts and tools in large drawers in the back of most stores.
The company sells an even larger number of parts online.

Tools

Harbor Freight
Harbor Freight sells inexpensive tools in retail stores and online.
Its online division ships orders quickly, and the company has a
large selection of hard-to-find tools at prices that make owner-
ship of an infrequently used item possible.

Online Auctions

Everyday, millions of people list, buy, and sell goods and services
of all types on online auctions. The world's largest online auction
and shopping Web site is eBay. Because of the huge volume of
business conducted online, it is often possible to purchase rather
difficult-to-find items including chemicals, tools, and parts of all
sorts. I have had both good and bad experiences with the quality
of items I've purchased, but I've never been out-and-out
defrauded by a seller.

Scrounging

Scrounging for parts on the surplus market, whether in the form
of a swap meet, storefront, auction, or junkyard, can provide
some superb deals. Sometimes the only way to get a needed part is
to visit a vast number of weird or dirty piles of mechanica arcana.
Foraging often pays off, although it takes knowledge to get the
parts you need at the price you want.

In my position as contributing editor at *Make* magazine, I have
designed and built a great number of interesting projects. I don't
do this in a vacuum; there is a big community of helpful makers
who willingly share their expertise. It's been my privilege to get to
know some of the best parts scroungers on the planet. Armed
with years of experience, they can quickly separate the worthwhile

from the worthless, and they know at a glance which equipment and parts are valuable to whom.

Among my good friends is Mr. Jalopy. As his nickname implies, Mr. Jalopy is a speed lover with an automotive bent. An inhabitant of the Golden Third, he's built a lot of edgy, cool stuff, ranging from his eponymous hot rods to the world's largest iPod. He writes several blogs, and, like me, he is a contributing

Mr. Jalopy's Top Tips for Garage Sale Scrounging

- Be Early—Especially after the advent of eBay, the competition for "stuff" is fierce. The key to finding the best stuff is hitting as many garage sales as possible before 9 A.M.
- Match Target Area to Interests—If you are interested in buying machine tools and car parts, then you will search garage sales in different areas than if you are interested in vintage handbags.
- Map It Out—Make photocopies of your local road map and mark your targeted garage sales as culled from online and newspaper sources. Put a star next to those that sound particularly compelling.
- See the One in a Million—A solid garage sale can have 10,000 items, but we are looking for only the choicest articles from that mass. It helps to have a few key items that you can ask about, for example, "Do you have any decoys of mallard ducks or torque wrenches?"
- Be Respectful—Don't talk bad about people's stuff. Be a respectful guest. Negotiate with vigor but never grind people on the price.

editor for *Make* magazine. An expert on Obtainium, he has made a science out of what he has learned from visiting garage sales and swap meets for more than 10 years. His methods enable him to pick up loads of valuable raw materials along with the occasional out-and-out treasure.

Another scrounging tip is to look closely at working parts still inhabiting a nonworking whole. Consider an old garden tractor as an example. When the cutting deck rusts through, it is no longer useful for yard chores. Most people will be glad to simply give it away. But the guts of the machine are filled with still-functional parts, easily available to the creative scrounger. A day spent with a cutting torch, gear puller, and pry bar can provide enough wheels, bolts, bearings, shafts, belts, gears, and what-not to fill several parts bins.

7

The Thundring Voice

Let me still heare the Cannons thundring voice,
In terror threaten ruine; that sweet noyse
Rings in my eares more pleasing than the sound
Of any Musickes consort can bee found.

—"THE SCOTTISH SOVLDIER,"
GEORGE LAUDER, 1623

This chapter is bound to be controversial. The Milquetoasts of the world will find it intimidating or unsettling, while the outlier Evel Knievels and Jack Parsons could possibly find it a bit dull. But this book isn't for them anyway. It's for the sensible yet adventurous readers: the men and women within the Golden Third. For them, playing with fire, literally, can be exciting and enlightening. We'll begin this adventurous chapter by getting to know the most important chemical discovery in the history of the world: black powder.

The Fire Drug

It transformed the fate of nations. It changed the way wars were fought. It made weak countries strong and strong kingdoms weak. It ended the Middle Ages and started the Renaissance off with a bang. Its gush of hot, expanding gas blew away feudalism, as chain-mailed knights on horseback, invulnerable to handheld weapons and arrows, could no longer bully and dominate the commoners living in their fiefdoms. In my estimation, black powder (or gunpowder—the terms are synonymous) is the most significant chemical compound mankind has ever developed.

Because gunpowder was cheap and relatively simple to make, compared to fashioning armor, it became the great equalizer among those who fought. It led to the supremacy of technology over arm strength, making engineers and scientists more important than knights and ninjas. Francis Bacon, English statesman, essayist, and philosopher, wrote that gunpowder "changed the whole face and state of things throughout the world; so much that no empire, no sect, no star seems to have exerted greater power and influence in human affairs than this."

For a thousand years, black powder was the only human-made propellant and explosive in existence, making it the most powerful, deadly, entertaining, and politically affecting chemical on earth.

Despite its essential simplicity, compounding black powder is not for everybody. While relatively tame compared to its high-energy cousins flash powder and smokeless powder, black powder has more than enough brisance to blow off valuable body parts if handled carelessly. But after reading several books dealing with the subject, I decided I couldn't intimately understand the stuff until I made it myself. Intellectually stimulating and historically revealing, making black powder gave me a degree of insight into this most important discovery that reading about it never could.

The first time I smelled the smoke and saw the fire issuing from the magic powder I'd compounded, I knew that it was something special. I'll remember the bang for the rest of my life.

The Chinese invented gunpowder, calling it the fire drug. The dates are unknown, but by the 10th century, gunpowder was in use for ceremonial and entertainment purposes, if not for warfare. Records tracking Chinese royalty recount an incident in which the Emperor Lizong's elderly mother, Kung Sheng, flew into an apoplectic frenzy due to a magic stunt gone wrong. Apparently a gunpowder rocket shaped like a rat flew wildly around the throne room, narrowly missing the dowager and causing much anxiety in the palace.

Gunpowder is composed of three ingredients: potassium nitrate (often called saltpeter), sulfur, and charcoal. The sulfur and charcoal were easy to find. But the saltpeter was sometimes a more difficult acquisition.

In parts of China, gunpowder makers could merely scoop up saltpeter that lay on the ground, the result of fermentation of soil and animal waste in the humid subtropical climate. Europeans living in a dryer, colder environment had to work much harder to get saltpeter. The early European method of obtaining potassium nitrate involved aggregating great heaps of rotting organic matter, especially that which contained high percentages of rotted meat and animal dung. "Petermen" would search out promising places to collect their smelly treasure. Abandoned outhouses and animal pens were especially prized. The petermen picked up and taste-tested handfuls of dirt. When they found a place that tasted right, they'd cart out the soil, boil it in vats, then evaporate the liquid and strain the slurry-like residue. The result was high-purity saltpeter.

Once you have the ingredients, you can't just shake these three chemicals in a jar. The ratio and manner in which they must be combined are precise and unforgiving. Mixed in the right proportion, the chemicals become what the Chinese called "magic black powder." Combine them incorrectly and you get a mound of unimpressive, dirty-looking powder.

The correct ratio is 75 percent saltpeter, 15 percent charcoal, and 10 percent sulfur. Each ingredient has a specific job to do in producing the desired chemical reaction. The charcoal is the fuel. Made correctly, charcoal is virtually pure carbon. Unlike other forms of pure carbon such as coal or diamonds, its structure is a chemical lattice, filled with microscopic pits and voids that are critical for rapid burning.

The saltpeter is what chemists call an oxidizer. It willingly gives up the oxygen locked within its chemical structure to a fuel. That allows for burning. Of course, charcoal burns with the oxygen available in the surrounding air. But if the oxygen for burning is supplied chemically, by mixing it with an oxidizer such as saltpeter, the reaction happens far faster and with greater gusto.

The sulfur, called brimstone by alchemists in the Middle Ages, plays a dual role: it facilitates detonation by lowering the temperature at which saltpeter ignites, and then it increases the speed and intensity of the ensuing chemical reaction. Propellant powders can be made without sulfur, but they are harder to ignite and they just wouldn't be gunpowder.

My experience in obtaining ingredients of sufficient purity and quantity wasn't trivial, but it wasn't overly difficult either. The list of vendors in chapter 6 will yield several sources.

Once I procured the saltpeter and sulfur, the final and most difficult step was obtaining charcoal. This runs counter to intuition since bags of charcoal briquettes are piled up by the front door of most grocery stores and gas stations. But that charcoal won't work, adulterated as it is by chemical binders and additives. No, in order to obtain the purity required, I had to make my own charcoal.

Wood roasted in the absence of air forms charcoal. I obtained all the charcoal I needed by wrapping small hunks of willow wood in airtight aluminum foil, then leaving them overnight mixed in the remains of the still-hot charcoal briquettes from a cookout.

Next came the most involved part of the process. It's called the precipitation reaction. The three chemicals turn into true black powder only when the saltpeter and sulfur are inserted at a minute level into the microscopic nooks and crannies on the surface of the charcoal particles. Squeezing the oxidizing saltpeter deep inside those pores takes a lot of time and a lot of patience. Commercial gunpowder makers typically use a device called a ball mill to accomplish the task.

A ball mill is a rotating cylinder containing nonsparking lead balls. The gunpowder's raw ingredients are placed inside the cylinder and then rotated slowly for hours. The balls grind the saltpeter, sulfur, and charcoal together until they are mixed at an almost molecular level of intimacy. The longer you mill, the better the explosive characteristics of the powder.

After careful measuring, mixing, and grinding, I removed my first-ever batch of powder from the ball mill cylinder. I knelt down on my driveway and placed a pile of powder, the size of a single Rice Krispy, on a piece of paper to test it. I lit the corner of the paper with a match. The closer the flame came to the small mound of powder, the farther back I stood, unsure of what would happen when the flame reached the powder. There was a flash of orange light, an audible fizzle, and then a great plume of black smoke. My heart lept; this stuff *worked*.

The Science of Black Powder

Black powder is actually medium gray. A pyrotechnic composition, it is highly combustible and when exposed to a flame has the wonderful ability to quickly change itself from a small amount of powdery solid into expanding volumes of hot gas.

Still, gunpowder is rated as a low explosive because the rate at which the exploding gas attains its maximum pressure is relatively leisurely. A chemist would tell you that compared to more energetic compounds such as dynamite and nitroglycerin, gunpowder actually deflagrates instead of explodes. Deflagration means that the speed at which the gas expands is subsonic. The more high-powered explosives unleash a supersonic detonation of far greater brisance.

In the instant it takes gunpowder to detonate, it increases in volume about 4,000 times. Put the black powder in a cannon and place a cannon ball in front of it, and the cannon ball will be pushed out with enormous force. The suddenly gigantic volume of gas forces the ball down the muzzle with enough momentum to travel hundreds, if not thousands of feet in free, ballistic flight.

The chemist's notation that describes the reaction is:

$$2KNO_3 + S + 3C = K_2S + N_2 + 3CO_2$$

Upon ignition, the trio of chemicals reacts and converts into potassium sulfate, nitrogen, carbon dioxide, and assorted other by-products. The potassium-based after-products appear as the dense white smoke characteristic of black powder.

Making Black Powder

As mentioned above, you simply can't take two parts of potassium nitrate, three parts of carbon, and one part of sulfur, shake them up in a jar, and obtain gunpowder.

I can't begin to tell you how many friends and business associates laughed, turned pale, and/or ran away when I explained that my book would contain information that allows readers to make their own gunpowder.

"That's illegal!" they shouted.

"That's what terrorists do!" they cried.

"That's too dangerous!" they warned.

I beg to differ. First of all, in many countries, possessing small amounts of black powder is not regulated or much cared about. However, various laws or ordinances may come into play, so it's advisable to check with local authorities prior to beginning. Second, in most places, it's far easier to buy large quantities of higher-power, higher-quality black powder at a sporting goods store than it is to make it yourself. And let's not even get into the subject of fertilizers and fuel oil. Which brings up the matter of making things of use to terrorists.

As far as the terrorist potential for homemade black powder, it's pretty minimal. Even underage delinquents have easier opportunities for finding materials with which to cause problems than to go through the rather long and demanding processes described here.

And as for danger, well, of course it's dangerous if you're not careful. But so is driving a car or mowing your lawn. (And, I repeat, things can and do go wrong, even for people who are being careful. There is no way to eliminate all risk.) But if you follow the procedures and limit quantities, you likely have bigger things to worry about in terms of safety than what's included here.

Safety

Let's tackle the safety issues head-on. Here is what you should and should not do.

First, don't make large quantities of black powder. Don't make even medium quantities. Limit batches to a maximum of one ounce or less of final product. Even that's quite a bit, because a mere fraction of that much high-quality black powder can maim or kill. Therefore, it must be stored safely and securely in a tightly capped container away from heat.

Never use more than 50 milligrams of any explosive powder (a pile smaller than half an aspirin tablet) in any single pyrotechnic device. This is probably a good time to discuss making M-80s or any other large firecracker. A true M-80 contains much more than 50 milligrams and is illegal in every state in the United States and most developed countries. The downside consequences of making and using such pyrotechnic devices (losing fingers, hands, eyesight, hearing, and possibly going to jail) far outweigh the brief instant of "fun" you get from the loud bang. Be smart and stay in the Golden Third.

You probably assume that your fuse will provide you with a specific margin of safety with respect to time. But in the DIY world, your powder could leak or spill, causing nearly instantaneous firing. Therefore, use safety equipment when making or using anything involving black powder. This includes safety glasses, a face shield, gloves, and a leather lab or welder's apron. Avoid wearing synthetic fabrics because they can melt when exposed to heat.

Don't substitute other materials or chemicals for the ones described. Powdered metals and chlorates are egregious examples of dangerous additions or substitutions.

That's the first group of common sense rules for playing with fire. If after you've digested those rules you are still committed to living dangerously but acting rationally, here is another group of practices designed to keep you alive and well and operating within the Golden Third.

Isolate your lab from everything else. Making gunpowder in the kitchen is a bad idea. If you have an accident, kitchens are expensive to repair.

Work in a well-ventilated area. Smoke and fumes are possible by-products of the process. The less you inhale, the better off you'll be.

Carefully prepare the area in which you make black powder. Keep the area clean and remove any spills immediately. Don't leave chemical containers uncovered on the workbench. Clean measuring equipment, mixing spoons, weighing containers, and so on immediately after using. Don't contaminate chemicals or mixtures by using dirty equipment.

Never place explosive or fast-burning material in any metal container. Exploding metal is shrapnel. Don't use a metal tube or pipe for any pyrotechnic purpose.

Never, ever grind mixtures of sulfur, saltpeter, and charcoal in a coffee or other high-speed grinder.

Smoking in the vicinity is strictly forbidden.

Sparks are a major safety concern because sparks trigger explosions. Static electricity, friction, and electrical contacts are sources of sparks in the workshop. Humidification and antistatic laundry spray inhibit static but may not prevent it completely. Be particularly careful to avoid sparks when screwing or unscrewing the lids of storage vessels containing black powder.

Steel against steel, steel against concrete, and certain types of stone hitting stone can create sparks. You can avoid sparks by using static-free plastic utensils for mixing and measuring. Similarly, use nonmetallic bowls when mixing chemicals and never point the open end of a mixing container toward yourself or another person.

Don't turn electrical devices on or off when explosive materials are exposed.

The work area must have a fire extinguisher and access to water for cleaning up spills or accidents.

If this list of safety precautions hasn't scared you off, it's time to get excited; you're following in the footsteps of some of the most important intellectual daredevils that science has ever known. Making your own black powder puts you in the rarefied company of such important historical figures as Joan of Arc, Roger Bacon, Mark the Greek, Lammot du Pont, Black Berthold, and Leonardo da Vinci.

The Basic Ingredients

You'll need three chemicals to make black powder: charcoal, sulfur, and potassium nitrate. See chapter 6 for information on where to purchase these materials. Instead of buying charcoal powder, consider making your own charcoal. Charcoal is easy and interesting to make.

How to Make Charcoal

As noted earlier, charcoal briquettes are not usable for black powder because they have additives. They're great for barbequing steak but lousy for making powder. Making your own charcoal is not difficult. Pure lump charcoal, or wood char, is nothing more than wood roasted in the absence of air. Without air, the wood doesn't oxidize or burn. Instead, it more or less bakes. The process removes water, oils, tar, and other volatiles and turns what's left into a dark, chalky residue of carbon.

After all the goo and moisture have been heated away, the remaining wood char weighs about 20 percent as much as the original hunk of wood. This is pure lump charcoal, and it burns hotter and more slowly than the original wood, no matter how well seasoned or dry it was.

Materials and Tools

- Several small chunks of wood (Willow is the traditional choice for gunpowder charcoal but poplar, alder, or cottonwood are also good choices. Avoid harder, denser woods.)
- Heavy-duty aluminum foil
- A medium sized nail
- Regular charcoal briquettes
- Grill

- Charcoal lighting method (e.g., charcoal-starting chimney or lighter fluid)
- Long-handled lighter or fireplace-style matches
- Grill tongs
- Mortar and pestle or coffee grinder (specially reserved for grinding individual chemicals)

Step 1. Wrap small chunks of wood in multiple layers of heavy-duty aluminum foil. Fold the edges tightly to make it airtight.

Step 2. Use the nail to poke a small hole, approximately $\frac{1}{8}$-inch in diameter, in the end of each piece of foil-wrapped wood. The hole is large enough to let steam escape, but small enough that the bulk of the wood inside the foil roasts without air but doesn't burn.

Step 3. Place regular charcoal briquettes in the bottom of a grill and light them. When the briquettes are covered with white ash, the coals are ready. (Incidentally, the white color comes from the addition of lime, which acts as a signal that the coals are ready. The lime is another reason you can't use charcoal briquettes as a black powder ingredient.) The number of briquettes you need depends on the amount of lump charcoal you require. A 5-ounce piece of wood becomes a 1-ounce piece of charcoal.

Step 4. Place the wood bundles in the grill and use the grill tongs to cover them with briquettes.

Step 5. Allow the briquettes to burn down to ash. This will take several hours. When the briquettes have burned away, remove the wood bundles with the tongs.

Step 6. When cool, unwrap the charcoal and grind it into a fine powder using a mortar and pestle or a coffee grinder. Your charcoal is ready to be used to make black powder.

Grinding charcoal

The Field Expedient Method of Making Black Powder

I first attempted to make black powder using a ball mill, which is how black powder is often made commercially.

A ball mill grinds materials into extremely fine powder for use in paints, pyrotechnics, and ceramics. This machine is basically a rotating chamber filled with the material to be ground and milling media. The milling media, in this case lead balls, pounds the material as it rotates, reducing it to a powder as it thoroughly mixes all the materials inside the chamber.

As the mill grinds the charcoal, sulfur, and saltpeter together, the pores on the surface of the charcoal particles are filled with even finer particles of sulfur and saltpeter. The longer the material turns in the ball mill, the finer the powder becomes.

But ball milling is difficult and time consuming. Furthermore, it can be dangerous, as a spark could ignite the contents in the mill. The ball milling method of making black powder in very small quantities doesn't scare me, but making larger quantities may cross the line from the Golden Third into outlier territory.

After much time spent attempting to design and build ball mills with the correct rotational speeds and capacities, I came across a method I like even better. A straight-forward, milling-free method of making black powder was first detailed in a 1967 U.S. Army document called "The Frankfort Arsenal Memorandum Report M67-16-1." This procedure is commonly known among pyrotechnic enthusiasts as the "field expedient method for the preparation of black powder." Truly, making your own powder from hardware store materials is James Bond or *MacGyver* stuff, but don't get carried away, figuratively or literally!

The field expedient method foregoes the need for the milling operation by exploiting potassium nitrate's extraordinary solubility in hot water. In this procedure, potassium nitrate, sulfur, and charcoal are dissolved in hot water, becoming intimately mixed. The resulting solution is dumped into a larger quantity of alcohol. The alcohol drastically reduces the water's ability to hold the saltpeter in solution, causing a large amount of high-quality black powder to precipitate out. The powder is obtained and dried using some simple screening techniques. Once dry, the powder is small grained and of good quality.

Before you begin making black powder, it's a good time to take stock. You now understand the solubility characteristics of the chemicals. You know how to make charcoal, and know from chapter 6 where to obtain potassium nitrate and sulfur. The final step is to make real black powder—the devil's distillate, the fire drug, artificial fire, or any of the many nicknames by which this most important of chemical discoveries was known.

Perform the following project in a well-ventilated work area such as a garage or outdoor workshop. Chapter 5 includes a long list of important considerations, including why your kitchen isn't a good spot for making black powder. This would be a good time to review that information.

Materials and Tools

- Safety glasses and a face shield
- Leather lab or welder's apron
- Fire extinguisher
- 75.0 grams of technical grade or better potassium nitrate (saltpeter)
- 15.0 grams of ground pure lump charcoal or wood char
- 10.0 grams of technical grade or better sulfur
- Mortar and pestle or coffee grinder (used only for grinding individual chemicals, not for food use)
- Wooden spoon or spatula
- 4- to 6-quart pot
- ½ cup water, divided in half

- Electric hot plate
- 1 pint high-proof vodka or 70 percent isopropyl alcohol cooled overnight in a refrigerator
- (2) 2-gallon plastic pails
- 2-by-2-foot square of cotton cloth such as an old T-shirt or diaper
- Tape
- Kitchen strainer with medium to fine holes
- Clean plastic container
- Cookie sheet
- Nonglass storage container with tight fitting lid such as plastic Tupperware

Step 1. If you have granulated, or "prilled," potassium nitrate, place it in a coffee grinder and reduce it to a fine, dust-like consistency. I tried to do this with a mortar and pestle. It's tough on your hands and difficult to grind the powder fine enough. But a few seconds in an inexpensive coffee grinder does the trick.

> Note 1: The coffee grinder is for grinding individual chemicals only. *Never use it for milling combined ingredients,* as this is a very serious explosion hazard.

> Note 2: Most chemists would recommend that you not use your regular coffee grinder for grinding chemicals. According to the Material Safety Data Sheet, ingestion of potassium nitrate, sulfur, and other chemicals is bad for you. It will be difficult to clean your regular grinder well enough to remove all the dust. Since coffee grinders can be relatively inexpensive, I recommend you bite the bullet and procure one just for this purpose.

Step 2. Use the wooden spoon or spatula to thoroughly mix the dry chemicals together in the large pot. Add ¼ cup of water and mix thoroughly.

Step 3. Put on your safety glasses, face shield, and apron. Place the pot on the hot plate and heat on medium. Add another ¼ cup water and mix thoroughly. Continue to heat until small bubbles begin to form. As you heat, make sure all of the mixture stays wet. Don't let any of the mixture dry out, for instance as residue on the sides of the pot, because it may ignite.

Dissolving ingredients

Step 4. Pour the chilled alcohol into one of the plastic pails. Remove the pot from the heat and pour the mixture quickly into the alcohol while vigorously stirring the alcohol with the wooden spoon. Let the mixture stand for five minutes. The black powder will precipitate out of the liquid.

Step 5. Affix the cotton cloth firmly to the top of the second pail with tape. Strain the liquid from the first pail through the cloth to obtain the black powder precipitate.

Step 6. Wrap the cloth around the black powder and squeeze to remove as much excess liquid as possible.

Cloth filter

Removing water

Step 7. With the wooden spoon or spatula, scrape the wet powder from the cloth. Place it on a dry flat surface such as a cookie sheet and let it dry until it is slightly damp.

Step 8. Place the kitchen strainer over the clean plastic container and press the damp powder through the strainer screen and into the container. This step is called corning or granulating, and it results in increased speed and performance of the powder.

Damp black powder

Screen

Corning black powder

Step 9. Spread the black powder granules into a ½-inch thick layer on a clean, flat surface such as a cookie sheet and allow it to dry in the sun, under a lamp (use a 60-watt bulb, but not too close), or near a radiator. Preferably, the mixture should dry within one hour. The longer the drying period, the less powerful the powder will be.

Caution: Remove the black powder and transfer it to a safe (non-glass) storage container as soon as the granules are dry. Your field-expediently prepared black powder is ready for use!

Troubleshooting Black Powder

Making black powder is conceptually easy, but you'll find that it really does require a lot of care and precision and takes time and repetition to master. It is possible, even probable, that the first batch you make may not work very well.

When the performance of my initial attempts at manufacturing black powder was disappointing, I turned to Ian von Maltitz for advice. Von Maltitz has been interested in fireworks and pyrotechnics since he was five years old. Well respected within the pyrotechnics community, he is among the world's most knowledgeable experts regarding gunpowder production. His book *Black Powder Manufacturing, Testing, and Optimizing* is the information source serious black powder makers rely upon.

Ian von Maltitz's Top Tips for Making Black Powder

- Your raw ingredients must be of good quality. Old, cheap, or adulterated chemicals will result in disappointment. Poor-quality ingredients are the number-one cause of poor performance.
- The sulfur and the potassium nitrate must be reasonably pure, but technical grades of these chemicals are usually adequate. However, it is the quality of the charcoal that makes or breaks black powder. Use the best-quality charcoal available. Make your own charcoal from willow or cottonwood if possible.
- There are two keys to fast-burning powder. The first key is to grind the materials as finely as possible. The second is to be certain the ingredients are mixed very, very well.

- Measure the ratio of chemicals within the composition carefully. Use a quality scale and be certain that spillage and overlooked ingredients left in spoons or dishes do not ruin your result.
- For safety, be extremely vigilant about controlling static electricity in the shop, and if you use a ball mill, be certain to use the correct type of (nonsparking) grinding media.
- When working with black powder, one method or way of doing something is not *safer* than another. Instead it is merely *less dangerous*. Working with black powder always contains some degree of danger.

Please keep von Maltitz's words in mind as you read what follows. In the next chapter we're literally playing with fire.

8

Playing with Fire

The Foreman's name was Pat McGann,
B'gosh he was an awful man;
One day a premature blast went off,
And a mile in the sky went big Jim Gough.
—FROM THE FOLKSONG
"DRILL, YE TARRIERS, DRILL"

In this chapter I cover two important questions. The first is, how do we control the interesting but dodgy stuff we're making, igniting it when and where we want it?

Second, once we've made black powder and the fuses to control it, what shall we do with it? A fine line separates projects compatible with the Golden Third and those that go beyond into outlier territory. The projects in this chapter, including rockets, cannons, and smoke bombs, do fall squarely within the bounds of reason if you combine the safety guidelines presented here with your own common sense.

Black powder, stand-up comedy, and jazz have something in common: timing is everything. Fuses are the devices by which you make things happen when and where you choose.

Fuses are themselves fascinating little constructions. If made well, they burn with gusto, yet possess the critical precision that

69

people bet their safety upon. And as they burn, fuses crack and sparkle, spit off chunks of fire, and emit roils of smoke, while making delightful "ffffffft" sounds. Fuses are indeed interesting little pyrotechnic devices, and understanding them is a worthwhile thing for any connoisseur of dangerous living.

Fuses are important because they have to catch fire reliably, stay ignited, and burn at a consistent rate, lest your experiments be plagued with misfires, duds, or worse.

Newspapers and magazines of the 19th and early 20th century are filled with articles about miners blown up by premature explosions. Life in the mines was hard, made harder by the danger of sudden and violent death.

Many of the news accounts of the incidents are extremely graphic. For example, under the headline "Terrific Premature Blast on the West Side" is this story lead: "George Smith, three-score years old, and expert in the use of high explosives was atomized shortly before 7 o'clock . . . Smith's comrades hunted for his remains. About 300 pieces which were collected weighted little more than 40 pounds so that the bulk of his body was reduced to mere shreds which could not be recognized." (*New York Times*, February 22, 1889)

The example of poor George Smith is provided not for its lurid details, but because it graphically shows how important a good fuse is to those who work with pyrotechnics.

Fuse Basics

The most basic type is known as a burning fuse. Invented in China at the same time as gunpowder, this type of fuse may be nothing more than a strip of paper dusted with black powder. But consistent performance of a crudely made fuse is not something you can depend on. When you're working with pyrotechnics, you must have something that is dependable and trustworthy.

Commercially made fuse offers excellent consistency and reliability. Commercial fuse is cheap, easy to obtain, and comes in a wide variety of diameters and colors. (You can find multiple fuse suppliers in chapter 6.) The best type of commercial fuse for the projects in this book is called Visco or safety fuse. Visco fuse is reliable and even burns in water. It consists of a center core of commercially made black powder wrapped in multiple jackets of wound cotton thread. The cotton is painted with colored nitrocellulose lacquer that makes it stiff and waterproof. Visco is manufactured under controlled conditions so that it burns at a consistent rate. Standard cannon fuse burns approximately one inch every two seconds, although you may find faster and slower burning varieties as well. Knowing this, you can cut a fuse to a length that gives you enough time to reach a safe place after lighting.

Be aware that Visco (and all other fuse types for that matter) burn vigorously once lit, spitting off very hot hunks of matter. Therefore, use gloves and long-handled matches or a barbeque-style lighter to ignite a fuse.

How to Use Commercially Manufactured Fuse

Fuse material is hygroscopic, meaning it will absorb moisture, which degrades its performance. It's best to store fuses in an airtight, sealed container, such as a plastic bottle with a screw-top lid or Tupperware.

There are different types of fuses for different applications. Some fuses are designed for pyrotechnics, some for cannons, and some work in multiple applications. The blackmatch fuse described here is a basic, utilitarian type, suitable for all the projects described in this book.

Match Fuses

In the 19th century, before Visco, safety fuse, detonation cord, thermalite, and other modern methods of initiating an ignition sequence, there were old-fashioned "match fuses." Historically, a "match" was a length of flammable cord or string saturated with chemicals. The speed with which the match burned depended upon the particular chemicals in which it was steeped. There were three general varieties: slowmatch, blackmatch, and quickmatch.

Slowmatch is a slow-burning fuse made from natural fiber rope that's been impregnated with an oxidizer chemical, such as saltpeter. The oxidizer particles that saturate the cotton or hemp rope fibers make the fuse suitable for manually lit shooting devices such as matchlock guns and black-powder cannons. This stuff burns very slowly. An advancing flame front consumes the combustible rope fibers at a leisurely four or five inches an hour. The punks (bamboo sticks covered with brown coating and saturated with nitrate) used for lighting consumer fireworks on the Fourth of July are a type of slowmatch.

One step up the fuse complexity scale is blackmatch. Blackmatch is similar to slowmatch except the surface of the cotton string is coated with a black powder paste. Blackmatch fuse burns hotter and much faster than slowmatch, spitting out a great deal of sparks as it burns. It's messy to make and use, but it serves its purpose quite well. I'll show you how to make your own blackmatch later in this chapter.

Finally, we'll look at the match fuse called quickmatch. There isn't much physical difference between quickmatch and blackmatch. In fact quickmatch is simply blackmatch wrapped in a paper tube. But holy cow—that paper tube causes a huge difference in the way it performs.

Quickmatch burns incredibly fast. Burn rates in the neighborhood of 50 feet a second are not uncommon. A foot-long piece of quickmatch appears to burn almost instantaneously. Quickmatch is often used in professional fireworks displays to obtain (nearly) simultaneous ignition of physically separated pyrotechnical devices.

Making Blackmatch

Well-made blackmatch is a dependable fuse, suitable for use in the large number of edgework practices requiring time-delayed ignition. To be honest, there's no way homemade blackmatch can match the consistency and ease of use of commercially manufactured fuse such as Visco. On the other hand, I think a burning blackmatch fuse is a beautiful thing to observe. Moreover, a large component of the living dangerously philosophy entails making things from scratch and doing things yourself.

Materials and Tools

- Oven
- 2 tablespoons cornstarch
- Cookie sheet
- 3 plastic spoons
- Safety glasses or goggles
- Apron or lab coat
- Leather work gloves
- Scale accurate to ½ gram or better
- 10 grams of black powder
- Quart-size mixing container with snap-on lid
- Electric drill
- ⅛-inch drill bit
- Knife
- 1 ounce isopropyl alcohol (purchase in any drugstore)
- 3 ounces water
- Spray bottle
- Popsicle-sized wooden stir sticks
- Small plastic container (about 6 ounces) with snap-on lid
- 2 to 3 feet of 100 percent cotton string, about ⅛-inch in diameter
- Scissors
- Drying rack (like a coat rack)
- Airtight plastic container
- Long-handled lighter or fireplace-style matches

Step 1. Making Dextrin

Blackmatch is basically cotton string evenly coated with black powder. The challenge is to get the black powder to adhere evenly to the string. The key is to mix the black powder with a chemical binder called dextrin into a semiliquid paste or slurry with which you coat the string. When the slurry hardens, it adheres to the string, turning it into blackmatch.

To make dextrin, preheat your oven (you can use your kitchen oven to do this step) to 400° Fahrenheit. Sprinkle 2 tablespoons of cornstarch onto a cookie sheet. Bake the cornstarch for 2 hours, mixing it thoroughly with a plastic spoon every 15 minutes to keep it from scorching. When the cornstarch becomes golden brown, it is dextrin. If the cornstarch turns nearly black, it has scorched and should be discarded.

Leave the kitchen, proceed to your workshop, and don your safety glasses, apron, and work gloves.

Step 2. Mix

Place 1 gram of dextrin and black powder into the quart container, affix the lid, and shake vigorously to mix them thoroughly.

Step 3. Preparing the Coating Container

Drill an ⅛-inch hole in the lid of the (approximately) six-ounce plastic container. Lightly clean up the hole with a knife so the edge is clean.

Step 4. Preparing the Black Powder–Dextrin Slurry

Mix one ounce of isopropyl alcohol with three ounces of water in a spray bottle. The alcohol causes the black powder-dextrin mixture to cling more easily to the surface fibers of the cotton string.

Gently spray the black powder mixture with the water-alcohol mixture and stir thoroughly with a stir stick. Continue to spray and mix until the black powder-dextrin mixture is a slurry, or semiliquid, similar in consistency to runny oatmeal.

Pour the slurry into the small plastic container.

Step 5. Immersion

Place the string into the slurry and use 2 plastic spoons to coat the string thoroughly. Thread one end of the string through the hole in the container lid and then snap the lid onto the container.

Blackmatch string and black powder slurry

Step 6. Coating the Fuse

Leave the string in the slurry for two minutes to let the cotton absorb the mixture. Slowly pull the string out through the hole in the container. If you've done the task correctly, the string will be evenly coated with the black powder slurry.

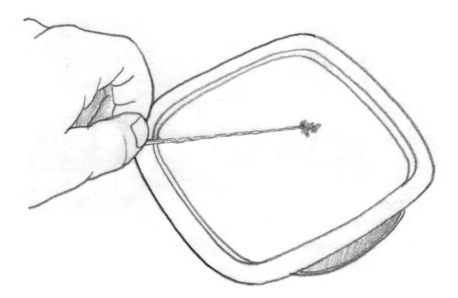

Blackmatch coating

Step 7. Drying the Fuse

Cut the string into 12-inch-long lengths. Hang the string fuse in a safe and undisturbed location until it is dry and crispy. I use a simple wooden board with several nails, but a coat rack also will do the trick. I merely loop the wet fuse over the nails and let them dry undisturbed. Of course, drying time will vary with weather conditions.

When dry, the blackmatch fuse can be cut with a knife to any length desired. Store the fuse carefully in an airtight plastic container.

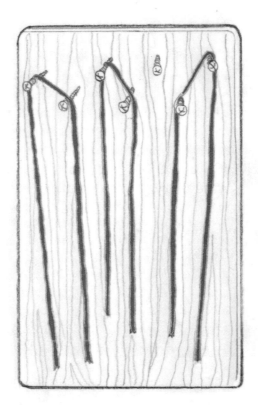

Blackmatch drying

Versatile and dependable, blackmatch can be used in a great variety of pyrotechnics and rocketry applications. You can use blackmatch for just about any fuse application, except for those that require the precise timing that commercially manufactured fuse provides. For example, it can be used to ignite the charge in a test mortar or eprouvette (see the following section), or it can be used to ignite a small castable propellant rocket engine (see chapter 9).

Making Quickmatch

Quickmatch is similar to blackmatch but it burns far, far faster. Burning rates of 100 feet per second are possible. Quickmatch is normally used to induce nearly instantaneous ignition in different parts of a fireworks tableau.

Quickmatch is just a blackmatch core inside a paper tube, or pipe, as it's called. When you light quickmatch, it undergoes what pyrotechnicians call *parallel burning*. The paper tube traps the hot gasses from the burning end of the fuse and reflects them back inside, causing the match and paper tube assembly to burn nearly instantaneously from end to end.

Making quickmatch is relatively simple and is interesting to experiment with because of its astounding burning rate. The burn rate is also reason to be very careful.

Materials and Tools

- Kraft paper 10 inches long by 4 inches wide
- 12-inch long, ⅜-inch diameter wooden dowel
- All-purpose white glue
- Homemade blackmatch (see previous project) or commercially manufactured Visco
- Airtight plastic container

Step 1. Roll a 10-inch-long piece of kraft paper around the dowel. Glue the seam so it forms a 10-inch-long, ⅜-inch-diameter paper tube. Take care to keep the glue off the dowel itself.

Step 2. Remove the dowel and let the tube dry.

Paper tube

Blackmatch

Quickmatch

Step 4. Insert the blackmatch fuse or Visco into the tube. (You can glue several paper tubes together if you need a longer quickmatch.) Crimp the ends of the tube around the blackmatch.

Step 5. Store carefully in an airtight plastic container until ready for use.

At this point you hopefully have a quantity of reasonable quality black powder, and you have the fuses with which to ignite it safely and reliably. What are you going to do with it?

There are only a few avenues you can take, essentially three variations on a theme. To understand the three, imagine you have a paper tube, caps that slide on the tube, a bottle of glue, and some black powder.

Variation One

If you place the black powder inside the paper tube, glue the caps on both ends, insert a fuse, and light it, you have a firecracker or bomb. The caps hold in the expanding gas until the tube ruptures with a bang.

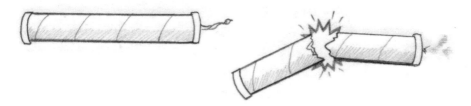

Variation Two

If you set the tube upright on the ground, glue on only one of the caps on the top, and ignite the powder inside, you have a rocket. The expanding gas exits the open end, causing the tube to move in the opposite direction.

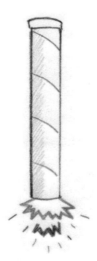

Variation Three

If you set the tube horizontally, glue on one cap and leave the other loose, then you have a cannon. This time, the expanding gas pushes on the sliding cap and turns it into a projectile.

In frontier America there was a tradition among blacksmiths called anvil shooting. On the Fourth of July, the village smithy amused the townsfolk by placing anvils on top of one another with a sizeable quantity of black powder filling a cavity cut into the steel surfaces between them. The smithy lit the blackmatch fuse and quickly took cover. The assembled crowd watched in breathless anticipation as the blackmatch burned down, spitting fire and sparks. Finally, the fuse contacted the black powder inside the anvil cavity. With a boom and a cloud of smoke, the top anvil would fly high into the air and then return to the ground, burying itself deep in the ground with an earth-shaking thud.

Anvil shooting is a testament to the considerable power locked within the chemical structure of black powder, as well as to the risk-taking proclivity of pioneer America. But while it sounds like a great show, the idea of a 64-pound anvil falling unguided from a height of 40 feet onto a field full of Fourth of July celebrants is probably nuts. So, no anvil shooting instructions here.

But don't worry, here is a project that utilizes the wonderful stuff we just made and does it a bit more responsibly. It's called the eprouvette.

An eprouvette is a small cannon. While its primary purpose is to evaluate the explosive capacity of a batch of black powder, it's more than a mere science experiment. It puts on a pretty good show, shooting small projectiles with great vigor and a visible energy that delights those who see it. This is a cool thing to make even if you use store-bought powder and fuses.

You'll discover something quite special about making your own cannon and gunpowder from household materials. If you desire more excitement, you can build on what you learned making the eprouvette and move (slowly and carefully) into making a larger cannon. However, I strongly recommend that you obtain expert advice before proceeding to larger devices.

Making an Eprouvette

Although it is primarily a simple cannon designed to test the strength of gunpowder, you can have a fine time shooting various small items from the barrel of an eprouvette. In practice, the powder maker places a carefully weighed quantity of black powder inside the firing chamber of the device. A carefully weighed projectile is placed on top of the charge. The charge is then ignited. The maker measures the distance the projectile travels. If it flies as far or farther than store-bought black powder, it's great stuff. Believe me, it's hard to make powder that good at home. If your powder is in the same ballpark, then good job! But if all you get is smoke and sputter, head back to your lab to try again. Remember, small differences in technique yield large differences in results.

Materials and Tools

- Scissors
- Kraft paper
- All-purpose white glue
- ⅜-inch diameter dowel, 1 foot long
- Knife
- Hacksaw
- 1 piece of 2 by 4-inch scrap lumber, about 4 inches long (The exact size doesn't really matter.)
- Homemade blackmatch or commercially manufactured Visco fuse. (A 3-inch, 6-second length of Visco fuse is ideal. If you're using homemade blackmatch, then test your fuse first to see how fast it burns, and use enough to get a 6-second burn.)
- Electric drill
- Drill bit slightly larger than the diameter of the fuse
- Ruler
- 50 milligrams black powder
- Test projectile (A small wooden plug cut from the 3/8-inch diameter dowel works well.)
- Safety glasses
- Lab apron
- Fire extinguisher

Step 1. Cut a sheet of kraft paper 6½ inches by 13 inches. Evenly apply a very thin coat of white glue to one side of the paper, except where the paper will touch the dowel. Wrap the 6½-inch side of the paper tightly around the ⅜-inch-diameter dowel. Use firm, even pressure to wind the paper around the dowel. Remove the dowel.

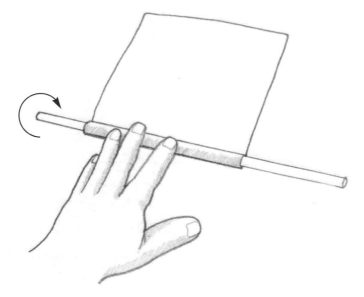

Eprouvette tube rolling

Let the glue dry and square off the ends of the tube with a knife or fine-bladed saw.

Step 2. Using a hacksaw, cut a 1-inch-long, ⅜-inch-diameter plug from the dowel you used in step 1. Apply glue to the dowel and insert it into one end of the tube. This forms the bottom of the eprouvette. Set it aside to dry. This paper tube and wooden dowel assembly is the cannon barrel of the eprouvette.

Step 3. Drill a fuse hole just above the wooden plug that is now inside the tube.

Step 4. Measure the outside diameter of the paper tube. In the center of the wooden block, drill a ½-inch-deep hole slightly larger than the tube diameter you just measured.

Step 5. Glue the tube into the wooden block and let dry.

Step 6. Cut a 1-inch-long piece of the ⅜-inch-diameter wooden dowel. This will be the test projectile.

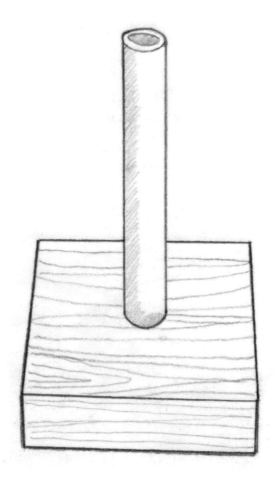

Eprouvette completed

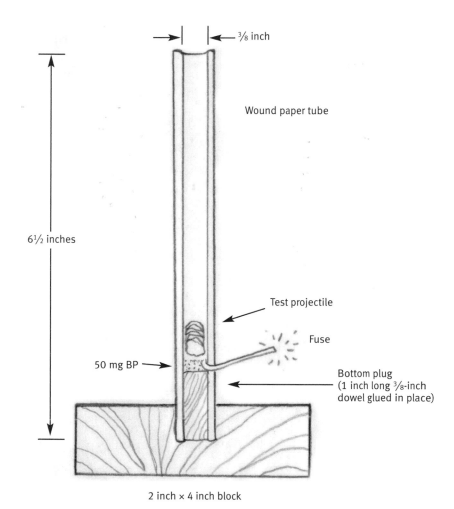

³⁄₈ inch

Wound paper tube

6¹⁄₂ inches

Test projectile

Fuse

50 mg BP

Bottom plug
(1 inch long ³⁄₈-inch
dowel glued in place)

2 inch × 4 inch block

Eprouvette cutaway diagram

Step 7. Firing

Place 50 milligrams of black powder inside the eprouvette. Fifty milligrams is a bit less than half an aspirin tablet. Insert the fuse into the fuse hole. Make sure the fuse length provides enough burn

time for you to get away safely after
lighting it. Insert the test projectile
and push it down into the tube until
it is just above the fuse and powder.
The test projectile should slide easily into the eprouvette.

50 mg black powder

Don safety glasses and lab
apron, and locate the fire extinguisher, 'cause it's powder-testing time!

Before you test your black powder, inspect your test area. As
the directions on every package of firecrackers warns:

> Use only under close adult supervision. For outdoor use only. Place on
> ground. Do not hold in hand or throw firecrackers. Light fuse and get
> away. Never attempt to relight a fuse. Never attempt to light firecrackers
> in a closed container. Never carry firecrackers in clothing.

I believe that much of the art in the art of living dangerously
is simply understanding which warnings really require your attention. The firecracker warning is one of those.

Speaking from experience, the lifting power between batches
of homemade black powder can differ enormously, because relatively small differences in procedure cause large differences in
performance. One batch may send the test projectile out of sight
while the next barely forces the projectile from the muzzle.

Note the altitude attained by the test projectile. You may have
to test several times to obtain reliable data. Once you know how
powerful your black powder is, you can compare it with the results
obtained from commercially made black powder.

There are myriad small variations you can attempt while stay-
ing fairly close to the cannon theme and maintaining a reasonable
risk profile. Earlier I discussed how bombs, rockets, and cannons
are variations on the way caps and tubes are arranged. In the next
chapter, we'll rearrange the caps and tubes and move from cannon to rockets.

9

The Inner MacGyver

An average person sees a hardware store as a place to buy some lag bolts and turpentine or to get window screens repaired. But a person who understands the art of living dangerously sees a hardware store as a place with immense creative potential.

In this section, I provide the information needed to walk into a store or two and come out with everything required to build a rocket. And I don't mean a toy rocket or water-powered rocket, but an honest-to-goodness, fuel-oxidizer-powered, solid-fuel rocket. As much as any project in this book, making homemade rockets from hardware store materials brings out my inner Mac-Gyver. Read and understand what follows, and you'll have the ability and knowledge to build a powerful rocket, similar in principle and construction to the space shuttle's solid rocket booster, just much smaller. It's a bit dangerous, of course, but boy is it artful!

Now, it's no secret that many stores sell amateur rocketry kits, including motors that may be larger and more powerful than what is described here. If you buy a model rocket kit and premade motor, there is little question that the rocket will fly the first time you attempt it, if you follow the directions carefully. It's easy. But it's not terribly exciting, and it's not living dangerously. It's a bit like buying a paint-by-numbers set: fun, perhaps, but not particularly creative. Making your rocket from scratch is a far more meaningful and, dare I say, thrilling experience.

In the past, many top rocket engineers and scientists began their careers without having formal degrees. In fact, much of the current body of knowledge concerning rocket science was developed from a knowledge base built from mid-20th-century amateur experiments. Consider the contributions of nondegreed rocket scientists Wernher von Braun, Hermann Oberth, and, of course, Jack Parsons. Their work led to tremendous advancements in the field of rocketry.

Given the current restrictive environment, could such outsiders make similar contributions in today's world? Between homeland security, our litigious society, and ever more restrictive rules and laws on the books, it seems unlikely.

Former NASA engineer Homer Hickam wrote a well-known book called *Rocket Boys,* and from it came a popular movie called *October Sky.* This was the story of four teenage boys from the backwater town of Coalwood, West Virginia, who designed and built a powerful rocket that flew nearly six miles into the sky. In one scene, Hickam and his friends started building their rocket engine by tamping down gunpowder in a metal pipe. This was not a good idea—and something far worse than property damage could have resulted. But the boys learned from their experience and eventually figured out the difference between building rocket engines and building pipe bombs.

Apparently Hickam's experience was not peculiar to him and his friends. People, especially teenage boys, went rocket-crazy in the late 1950s, their interest spurred to stratospheric levels during the patriotic frenzy caused by the first satellites launched into outer space. In those early days, thousands of young people attempted to build rockets. Few had the training and experience required to make rocket engines safely. Because accurate and reliable information was so difficult to obtain, many experimenters wound up building what in reality were pipe bombs.

And building pipe bombs, whether on purpose or inadvertently, is always a very bad thing. I have met individuals who lost fingers and other body parts due to rocket motor and other pyrotechnic mishaps. Homemade rockets and metal tubes don't mix.

But things have changed since the 1950s, and in very important ways. There are now a number of excellent books and manuals that explain in detail the procedures for safely building amateur rockets. And the Internet provides a highly connected and interactive forum for enthusiasts to learn directly from one another.

To me, making a rocket seemed to be a worthy goal for someone interested in edgework. I sought a small, easily made rocket that I could build without special tools, and with materials easily obtained from nonrestricted sources.

Early in my rocket-making quest, I cast widely for relevant information and perused several very interesting rocket-making instruction books. Among my favorites was Captain Bertrand R. Brinley's *Rocket Manual for Amateurs*, a paperback book written in 1960 by a man with the interesting-sounding job of U.S. Army liaison officer to civilian rocket enthusiasts. His book is jam-packed with formulas, recipes, drawings, tables, and information on all things rocket related.

I love many things about Brinley's book, especially its detailed diagrams of rocket nozzles and explosion bunkers and its exquisite cutaway drawings of rocket motors. But some of Brinley's advice, looked upon with contemporary eyes, is, well, a little questionable.

According to Brinley, the best and safest rocket engine is one made from a combination of zinc and sulfur. This combination of chemicals is known among rocketeers as *micrograin*. Micrograin is easy to make, requiring only the mixing of these two elements in powder form and tamping down the mixture with a wooden stick. Unfortunately, it's also very dangerous. Micrograin burns incredibly hot—necessitating a steel vessel, which can contain the highly exothermic reaction without melting or bursting. It also burns quickly, unquenchably, and uncontrollably. My first few experiments with micrograin rocket motors scared the heck out of me, quickly convincing me to abandon Brinley in favor of a more modern treatment of the subject.

Next, I considered making a compressed black powder rocket. Commercially made small rocket engines are typically made from black powder, which as you now know is quite straightforward to formulate. But to turn black powder into a rocket motor requires compressing it with literally tons of force so it forms a solid matrix that burns controllably and reliably. Without the right (expensive) equipment, I found homemade black powder motors to be a frustrating and probably dangerous endeavor.

Colleagues then pointed me in a new direction, one charted more than half a century ago by none other than Jack Parsons. It was Parsons who invented castable rocket fuel, that is, rocket fuel that starts as a soft, pliable material and slowly hardens, allowing it to be cast or molded into virtually any shape. Castable rocket fuel is easier to handle and is the basis for the modern solid fuel used by space shuttle booster rockets and other large launch vehicles.

On an amateur basis, I found that granulated sugar could be used to fuel a rocket motor. Basically, this method of rocket making involves melting a mixture of sugar and a chemical oxidizer over a hot plate and then pouring it into the rocket body, where it solidifies into a rock-hard casting that contains bounteous amounts of chemical energy. By mixing and melting potassium nitrate (saltpeter) and sugar in specific ratios, I could indeed obtain a relatively stable, safe-to-use rocket fuel capable of producing decent thrust and performance.

I experimented for days with different formulations of sugar and saltpeter, varying the ratio, the size of the particles, the shape of the motor, and everything else I could think of. The test bench performance of sugar motors was tantalizing, but I found it difficult to work with the molten sugar mixture since it quickly cooled into a brittle mass.

Then two acquaintances, very well known within the amateur rocket community, told me about a lesser-known method that utilizes a sugar-like powder called sorbitol. The results were amazing. Easy to work, pliable, and powerful, sorbitol–potassium nitrate motors were the answer I had been looking for.

Making a Rocket

Before you start, review the safety tips in chapter 5. Is your safety gear on? Then let's go!

Materials and Tools

- Safety glasses
- Face shield
- Lab apron
- Kraft paper (brown grocery bag paper works fine)
- ⅜-inch-diameter dowel rod
- All-purpose white glue
- Scissors
- 1 can of nonshrinking water putty (This is available in all hardware stores. A common brand name is Durhams.)
- Ruler
- Fine sandpaper
- Electric drill
- 5/32-inch drill bit
- Scale
- 13 grams potassium nitrate (This chemical is sometimes called saltpeter. Review chapter 6 for details on obtaining this material.)
- Coffee grinder, reserved for grinding chemicals
- 7 grams sorbitol (Available at health supplement stores and online. It comes as a powder, typically in a 1-pound bag. It looks and tastes similar to white sugar, but it is not quite as sweet.)
- Plastic mixing container with a tight-fitting lid
- A handful of large lead fishing weights or 50-caliber lead balls (Available at hunting and fishing stores. Do *not* substitute steel balls or marbles.)
- Electric skillet or pan with electric (*not* gas) hot plate

- Plastic spatula, the kind used in candy making
- Rubber mallet
- 6 D nail (called 6-penny)
- Thin wooden stick such as a bamboo shish-kebab skewer
- Clean, empty plastic bottle for storage
- Bucket of water (in case of a dud)
- Fire extinguisher

Step 1. Wind the Casing

Cut a sheet of kraft paper 4 inches by 10 inches. Wrap the 4-inch side of the paper tightly around the ⅜-inch dowel rod, and roll the paper around the dowel. Apply a very thin coat of white glue over one whole side of the paper. Roll the paper tube as tightly as possible. Remove the dowel and let dry.

After the glue dries, cut the tube into two 2-inch long tubes. Each paper tube will be suitable for a single-use rocket casing.

Step 2. Make the Rocket Nozzle

Reinsert the ⅜-inch dowel into the wound tube, leaving a 5/16-inch gap between the end of the tube and the dowel. Mix up a small amount of nonshrinking water putty according to the label directions. Press the putty firmly into the end of the wound tube. Slowly remove the dowel and set aside to dry. Repeat the process for the second motor.

When dry, use fine sandpaper to file the exposed surface of the nozzle flat and smooth. Mark the exact center of the nozzle and drill a 5/32-inch diameter hole through the nozzle.

Step 3. Prepare the Rocket Fuel

Rocket propellant isn't hard to make, nor are the raw materials difficult to get, but you do have to work carefully and precisely.

Only two ingredients are required: potassium nitrate and sorbitol.

Using your scale, measure 13 grams of potassium nitrate or saltpeter (KNO_3) and 7 grams of sorbitol. This makes enough propellant for a single 2-inch-long rocket motor. (In step 2 we made enough rocket casings for two rockets. However, for safety's sake, repeat this step to make the fuel for your second rocket instead of simply doubling the amounts listed.)

If your potassium nitrate comes in granulated or prilled form, reduce the prescribed amount to a fine powdery consistency in the coffee grinder. Again, as discussed in chapter 5, *never grind mixtures* of chemicals, only individual chemicals. Also, grind chemicals in a grinder different than your normal coffee grinder.

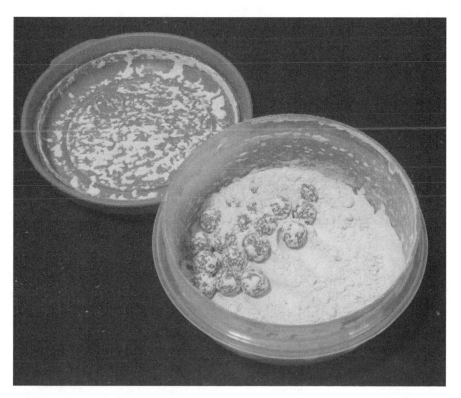

Rocket motor chemical mixing

Now put on your safety glasses and process the ingredients by placing them into a tightly covered plastic container with a handful of large, lead fishing weights or 50-caliber lead balls and shake for about half an hour. (Yes, I know that seems like a long time to shake something, but the longer you shake, the better the mixture works. You can watch a television show while you do this. It makes the time go faster.) Remember, you are working with a powerful mixture consisting of a fuel (sorbitol) and an oxidizer (potassium nitrate). Use only lead balls; steel ball bearings can spark as they collide. Handle everything carefully; keep it covered and away from ignition sources and children.

Melting rocket motor material

Now go *outdoors* to do what's next. (You may need to use an extension cord to use your electric skillet or hot plate outdoors.)

Put on your face shield over your safety glasses.

Heat an electric skillet or use a pan on a carefully controlled electrical (no open gas flame allowed) hot plate to 285° F. Add the rocket fuel mix to the pan. Stir continually with a hard plastic spatula or spoon. The mixture will soon melt into a jelly-like consistency. After a few minutes of heating and stirring, the raw materials will dissolve into a smooth paste, free of lumps and particles. Turn off the heat. The mixture will stay plastic and moldable long enough for you to make the rocket motor.

Step 4. Load the Rocket

Let the fuel cool until you can handle it with your fingers, and then carefully place the pliable rocket fuel mixture into the rocket casing. At intervals during the filling process, insert the ⅜-inch diameter dowel into the open end of the motor and tap the dowel with a few short, soft strokes with a rubber mallet to evenly fill the interior of the rocket motor with fuel. Fill the motor with fuel until the fuel is ⅜ inch from the top of the casing.

Rocket motor with castable load in place

Step 5. Core the Engines

In order to produce enough thrust to lift your rocket into the air, you'll need to make what rocket experts refer to as a core-burner. This style of rocket has an open center in the motor. This increases the area over which the rocket fuel burns.

You can use a drill bit to make an open core, but the castable propellant is very sticky and tends to gum up drill flutes. It is easier to simply insert a 6 D nail into the hole in the nozzle you drilled in step 2 and slowly push it through the fuel until the tip emerges on the other end. Carefully remove the nail by rotating it slowly so that the cylindrical hole remains intact.

Step 6. Seal the Open End

Seal the ⅜-inch gap at the open end with water putty. Let it dry completely.

Rocket motors

Step 7. Attach Balance Sticks

Glue a shish-kebab skewer or small wooden dowel to the rocket body with the stick extending backward from the nozzle. The length of the stick required depends on the weight of the rocket body. In general, choose a stick length such that the rocket body and stick will balance horizontally at a point one inch behind the nozzle.

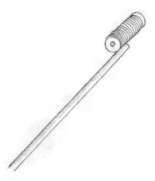

Rocket with balance stick

Set your motor aside in a safe and secure location. After the water putty seal and the glue that attaches the balance stick dry, your rocket is complete. The fuel will absorb moisture from the atmosphere, so use the motor as soon as possible or store it in a sealed plastic container such as a clean, empty plastic bottle.

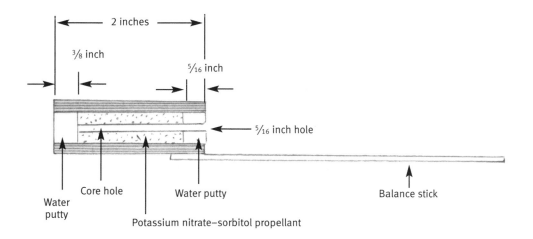

Rocket cutaway diagram

Step 8. Build a Launching Platform

Make a simple launching platform by attaching an 8-inch-long, ¾-inch- or 1-inch-diameter pipe to a scrap block of wood, using a floor flange and wood screws to support the pipe.

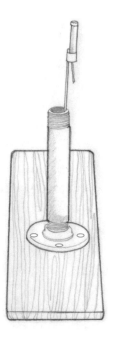

Rocket on launching platform

Step 9. Launch the Rocket

Launch rockets only in an open area, far from houses, people, and items that could be damaged from a falling rocket body, a launch pad explosion, or other unexpected occurrences.

There are several ways to ignite the rocket motor. The most straightforward is to insert a fuse directly into the rocket's opening, light it, and then retreat. The fuse may be made from homemade blackmatch or it may be commercially manufactured. (See chapter 6 for fuse vendors.)

Place the rocket inside the vertical pipe attached to a wood block. Make sure you're wearing safety glasses and light the rocket fuse with a long-handled match or lighter.

Optional. Build a Remote Ignition System

Another option is to build an electric igniter to ignite the fuse inserted into the nozzle. This allows adequate physical separation between the rocket and the rocketeer. A remote ignition system is easy to construct and eliminates some of the risks involved with rocket launching or fireworks. The igniter allows you to maintain a safe distance from things that explode or burn by eliminating the need to light a fuse and then retreat.

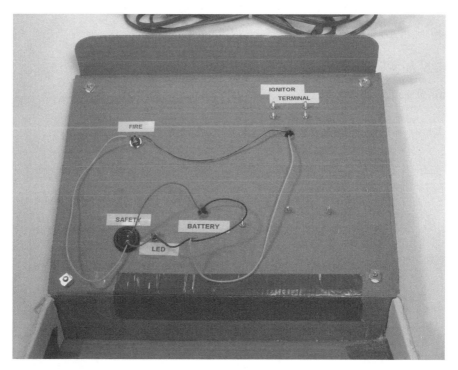

Remote igniter interior

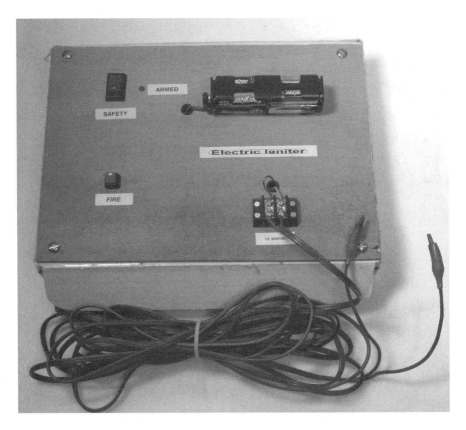

Remote igniter exterior

This system sends an electrical current through a thread-like metal ignition wire, heating it red-hot. The glowing wire ignites a short fuse that in turn ignites the rocket motor. The igniter consists of a number of AA batteries wired in series, a safety interlock switch, an LED to indicate "System Armed," and a firing pushbutton.

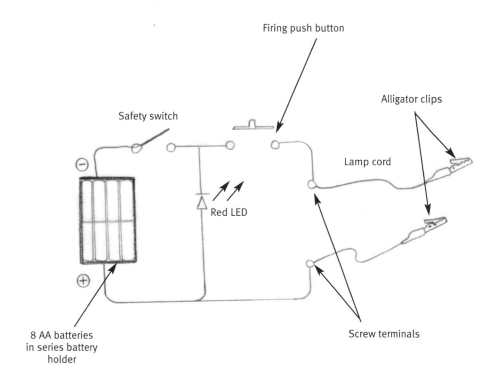

Firing push button

Alligator clips

Safety switch

Lamp cord

Red LED

8 AA batteries
in series battery
holder

Screw terminals

Remote ignition construction diagram

You can buy nichrome ignition wires online. To locate vendors, simply type "40 AWG nichrome" into any Internet search
engine. But it's easy enough to make your own ignition wires by
twisting together several steel wool fibers. The wire should be
about one inch long. Thread a sewing needle with the igniter wire
and then insert the igniter in the center of a short piece of Visco

or blackmatch fuse. Once the wire is inserted, connect one alligator clip from the electric igniter to each end of the thin wire as shown in the diagram.

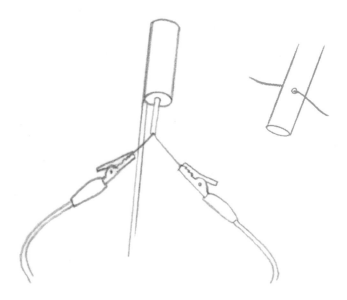

Rocket with fuse and igniter

Remove yourself to a safe distance of 20 feet or more from the rocket and turn the safety off. Perform a countdown and press the fire switch. The current will heat the wire and ignite the fuse, which in turn ignites the rocket motor.

After a successful liftoff, the rocket could reach altitudes of several hundred feet. Since the rocket's trajectory is difficult to predict, launch it only in appropriate areas (e.g., flat, open areas free from easily ignitable materials). Check local laws and make sure you have the landowner's permission to launch.

Troubleshooting

Making sobitol—potassium nitrate rockets is a straightforward process, but on occasion your rockets may fail to perform to expectations.

If you have a dud—that is, the rocket does not ignite—do not go near it for several minutes. Safely dispose of the rocket by soaking it in water then discarding it in a trashcan.

For the next rocket you attempt, be certain your chemicals are well mixed. Increase the mixing time for combining the chemicals in the shaking container.

Pay careful attention to the temperature setting on the electric skillet or hot plate. Melting at temperatures too high or too low can have detrimental effects on performance.

Make certain your chemicals are of adequate purity and your workspace is well maintained to avoid contamination.

If your rocket fires but will not lift off, increase the size of the nozzle opening, increase the size and depth of the core hole, or reduce the rocket's weight by using thinner paper and less glue when constructing the tube.

After several successful launches, you may be tempted to experiment with larger rocket engines. If bigger rockets with bigger motors call out to you, you're leaning to the right on the thrill curve.

Here's some advice from Dr. Terry McCreary, professor of analytical chemistry at Murray State University. He's also the author of *Experimental Composite Propellant*, the book many consider the definitive text on amateur high-power rocket-motor making. Dr. McCreary is well known among serious rocket makers and is a board member of the Tripoli Rocketry Association, the largest amateur high-power rocket making organization in the world. If your dreams include shooting for the stars, these guidelines are a good place to start.

Dr. Terry McCreary's Top Tips for Experimenting with Rockets

- Nothing is perfectly safe; think and work to reduce your hazards.
- Know the materials with which you work. Become familiar with the characteristics and limitations of the potassium nitrate and sorbitol.
- Duplicate the work of others (well-characterized formulas, techniques) before experimenting on your own.
- Consistency, consistency, consistency in everything. Take careful notes so you can reproduce your results faithfully later on.
- Attempt improvements incrementally. Small steps are usually more successful and less hazardous than big ones.
- However, scaling up doesn't always work; that is, making larger rockets by simply increasing ingredient amounts often does not produce the expected results.

Smoke Bombs

Smoke bombs and their military counterpart—smoke screens— have many uses. Smoke bombs are great fun when you want to conceal, divert, prank, or just make billowing clouds of smoke. Don't ask why this is fun; it just is.

You might ask whether making smoke screens is really a part of living dangerously. I maintain that it is. It involves fast-burning chemicals, fuels and oxidizers mixed together, and fire. Knowing how to make and use smoke bombs is a skill that has something so, let's say, *MacGyver*-ish about it.

Since they don't explode per se, smoke bombs aren't really bombs. But they do burn fast and vigorously, and while this project is mostly fun and games, it demands respect and caution.

Militarily speaking, smoke bombs or screens have been around since ancient times, but they did not become important as a military strategy until the 20th century. In both World Wars, great, wafting clouds of obscuring smoke provided protection to ships, allowing them to make evasive maneuvers out of sight of artillery observers. The descriptions of many battles are packed with references to the use of smoke screens. By all accounts it was an important tactical consideration, topmost in the minds of battle commanders.

Military tacticians list a number of uses for smoke in battle. Obviously a blanket of concealing smoke is useful for hiding the movements of troops and ships from the eyes of adversaries. But it has other uses. The Soviet army of the cold war era, for example, believed that blinding smoke, deliberately placed on enemy positions, was far more effective than using smoke to conceal their own positions. In fact, they found that their forces could reduce casualties by 90 percent by confusing enemy gunners and artilleries with clouds of blinding smoke.

Also, smoke bombs are extensively used by the military to identify friendly forces during heated battles when close air support and attack helicopters need visual clues to determine who is a friend and who is a foe.

Military smoke screens are typically made from some pretty dicey chemicals—white and red phosphorus, hexachloroethane, and fuel oil, for example. But you can whip up your own smoke screen for fun and excitement using safer and easy-to-find ingredients.

Making a Smoke Bomb

Here's how to construct a clever little smoke-making device that produces a wonderful plume of dense white smoke when ignited.

Materials and Tools

- 6 grams potassium nitrate (KNO_3): stump remover, technical grade, or better
- Scale
- Coffee grinder or mortar and pestle, necessary if using granulated or prilled KNO_3
- 4 grams powdered sugar
- .5 grams sodium bicarbonate (baking soda)
- Plastic mixing container with tight-fitting lid
- Safety glasses
- Electric skillet, or pot or heat-proof bowl and electric hot plate
- Plastic spatula
- Waxed paper
- Wooden or plastic mixing spoon
- Homemade blackmatch or commercially manufactured Visco fuse
- Scissors

Step 1. Prepare the Potassium Nitrate

If you're using prilled potassium nitrate (the type in stump remover) you'll need to reduce it to a powder. To do so, measure 6 grams of potassium nitrate (KNO_3) into a coffee grinder or mortar and pestle and grind it to a fine powder. (Remember, never grind mixtures of chemicals, only individual chemicals. Also, for grinding chemicals, only use a grinder separate from your normal coffee or spice grinder.)

Step 2. Measure

Measure the following using an accurate beam or electronic scale:
    6 grams of potassium nitrate
    4 grams of powdered sugar
    0.5 grams of sodium bicarbonate (baking soda)

Step 3. Combine Ingredients

Place the potassium nitrate, powdered sugar, and sodium bicarbonate in a closed container and shake until well mixed.

Step 4. Heat the Mixture

Put on your safety glasses and find a place outdoors for the following steps. Set an electric skillet for 285° F. Alternatively, you can use a pot or heat-proof bowl on a carefully controlled electrical (not open gas flame) hot plate heated to 285° F. Add the entire smoke screen mixture and stir continually with a plastic spatula. In one to two minutes, the mixture will soften and then melt into something that looks like runny peanut butter. When that happens, it's ready to be shaped or poured into a mold.

Heating the smoke bomb materials

Step 5. Pour onto Waxed Paper

Turn off the heat. (If the mixture overheats, it can ignite.) Carefully pour the hot, viscous mixture onto the waxed paper. Use a spoon to shape it into a rough triangular mound. Alternatively, you can pour it into a heat-proof candy mold.

Step 6. Insert Fuse

Cut a 2- to 3-inch length of Visco or blackmatch fuse, which should be sufficient to provide approximately four to six seconds of burn time. Insert the fuse into the mixture.

Step 7. Let Cool

The mixture will harden as it cools. After approximately 15 minutes the smoke bomb will be cool and ready for use.

Completed smoke bombs

Step 8. Using Your Smoke Bomb

It's simple to use your new smoke bomb. Just place it on the ground outside in a clear area, then light the fuse and get away fast.

A smoke bomb deserves your respect. Do not hold it in your hand when lighting it and do not throw it when it's burning. Smoke bombs are dangerous and should never be used indoors. Never attempt to relight a fuse, do not attempt to light a smoke bomb in a closed container, and don't carry a smoke bomb in your clothing.

In this and the previous two chapters, I outlined several projects that explored living dangerously by playing with fire. But living dangerously is more than this; it encompasses a wide range of experiences. In the next chapters, we explore concepts that go to the core of living dangerously. Some, like the chapters just presented, provide ideas and techniques for moving to the right on the risk-taking graph by making interesting, Big-T things. Others, like the next chapter, are purely experiential.

10

The Minor Vices

Sociologists consider such activities as drinking, smoking, lying, swearing, and gambling as the minor vices. But the word *minor* doesn't fit particularly well, for these vices are big business, big danger, big pleasure, and big money.

The minor vices are controversial. The world is divided into two camps: those who say you should never smoke, speed, gamble, or drink, and those who believe it's perfectly all right as long as your pursuits bother no one else.

One thing is certain: people pick and choose their vices, perceiving some as evil and others as fairly innocent fun. On television programs, people drink and gamble. In fact, there are entire television series on wine tasting and Texas Hold 'Em poker. But cigarette smoking has become increasingly rare on TV, and a curse-laden outburst can damage an actor's career. All of which make these taboos hard to understand; after all, too much swearing doesn't increase the likelihood of car crashes, and 18-year-olds can buy cigarettes in Vegas but are barred from the casino floors.

All of these vices are harmful if done too often and too recklessly. Done poorly, they cause relegation to society's bottom rung. American historian John C. Burnham labels the purposeful adoption of low-order social habits as an inverted parochialism. Hard-pressed individuals, it appears, may adopt bad habits in order to display a perverted sort of superiority over their intellectual or financial betters.

Becoming a lying, cigarette-addicted, alcoholic gambler who swears too much is certain evidence that one has failed at the art of living dangerously. Paradoxically, there are times and places where each minor vice can be pursued, when the act will do more good than harm. The trick is in knowing how, knowing when, and knowing where.

The minor vices are the Tabasco sauce and salt in the stew of human existence. In small amounts they add bite and depth to the flavor of life. But add too much and they'll ruin it completely.

Cigarette Smoking

In the middle of the 20th century, Camels and Lucky Strikes were nearly as important on the silver screen as Humphrey Bogart, Barbara Stanwyck, and Gregory Peck. The year 1949 saw the release of *Twelve O'Clock High*, a movie with one of the highest per-minute cigarette consumption rates of any movie ever made. The movie's publicity poster consists of a drawing of a flight-suited and goggled Peck, rock-jawed and determined-looking, taking one final drag on an unfiltered Chesterfield. Presumably, he is about to lead his squadron on an extremely dangerous bombing mission over well-defended ball bearing plants in Germany. Luckily for the men he commands, he looks ready to lead. The way he smokes says it all.

In a 1948 *Atlantic Monthly* article entitled "Smoke Without Fire," a London barrister with the delightfully theatrical name of Giles Playfair cataloged and analyzed 16 smoking-related communication gestures that stage and film actors used to express emotion. Here is an extract of Playfair's work, for your consideration and possible incorporation:

THE HISTRIONIC SMOKER'S COMPANION

To Show:	Do This:
Courage	Light a cigarette at every moment of danger. Male characters of the Tough American Hero school should take cigarette directly with mouth from a pack of cigarettes.
Anxiety	Take quick and frequent puffs at a cigarette while walking briskly. Discard half-finished cigarette and light another.
Caution	Light a cigarette, pipe, or cigar elaborately. Make as much as possible of extinguishing match by holding it up before one's eyes and regarding flame with interest before blowing it out.
Acute Distress	Crush out a half-smoked cigarette with awful finality.
Subtle Threat of Violence	Remove cigarette or cigar from corner of mouth with thumb and forefinger.
Passion or Desire	Put two cigarettes in mouth at same time. Light both; then, with possessive air, hand one of them to the adored one.

While the dangers of habitual smoking have been so well documented that they need no further discussion here, smoking a single cigarette, for a particular purpose, does qualify as a Golden Third activity.

The danger of smoking habitually is severe and extreme, but then again, doing any single act 30 times a day is bound to be bad for your health. But once in a while, done for the right purpose and in the right fashion, smoking a cigarette is OK. In fact, it's better then OK. This statement might surprise you. It might offend you. But hear me out before writing off my premise.

Psychologists tell us there are a number of reasons that people light a cigarette and inhale smoke into their lungs. It's available, pleasurable, rewarding, addictive, and so on. Cigarettes are good for killing time. Smokers can stand around doing nothing without being called loiterers. But all of these are insufficient, pedestrian even.

The only acceptable reason to smoke is to communicate. And it is a special type of communication that's connected with smoking; an intimate type when nuance and subtlety speak louder than words.

Warm, social communication—that's what cigarette smoking is useful for. Its only justification comes when it is used as a highly refined form of nonverbal personal interaction. When people come together with a cigarette in well-defined and carefully prescribed social conditions—a dark booth in the rear of restaurant, on a couch in an intimate corner at a party, or around a campfire on a chilly October evening—then comes the opportunity for close and intimate communications in smoky yet elegant gestures, acts, and words.

Consider the film noir movies of the 1940s and the symbolism of on-screen smoking. Clouds of smoke hover above Humphrey Bogart in *Casablanca*, Robert Mitchum in *Out of the Past* (which Roger Ebert called the greatest cigarette-smoking movie of all time), and Bette Davis in *Now, Voyager*. The smoke streams in and out of camera range; thin, roiling cones of white pushed from nostrils and pursed lips, gray wisps hanging top-frame in ghostly suspension, the tobacco fog eventually blown away by a cleansing breeze or camera fade.

In these movies, cigarettes are not just an actor's prop. The cigarettes are actors in and of themselves. They are important characters with vital roles that illustrate relationships among lovers, enemies, and friends. The way the actors smoke tells you wordlessly what's going on in their heads.

A shared smoke speaks eloquently. Bottom line, on those rare occasions when a cigarette can speak for you better than you can say it yourself, it's time to fire one up.

Safety

Before lighting up, consider that even very small amounts of smoke can be hazardous to your health.

Minimize the amount of smoke you inhale. Infrequent smokers will likely experience dizziness or nausea.

Do not smoke near others who object. Stand at least 50 feet from a doorway or window so that others are not subjected to secondhand smoke.

Obey all local laws and ordinances regarding smoking in public. Do not smoke around children.

How to Smoke

Choosing Your Cigarette

Since the artful and dangerously living smoker smokes only on rare occasions and only for a specific purpose, brand loyalty is not relevant. Since nonverbal communication is the primary reason for smoking, image is important. Choose a cigarette whose image is congruent with the situation. For example, if you are sharing a cigarette with a potential love interest, choose a brand whose image connotes personal intimacy.

European imports, especially those from France, often make good choices. There are particularly strong connections between sophisticated European intellectuals of the mid 20th century, such as Jean-Paul Sartre, Jean Cocteau, and Pablo Picasso, with the French brands Gauloises and Gitanes.

Alternatively, if you are enjoying a manly, same-sex-bonding experience, such as sitting by a campfire late at night with a beer, then an American cowboy—image cigarette or even a self-rolled cheroot is appropriate.

Lighting Up

The method with which you apply fire to the cigarette is a significant component of the communication. Again, the choice is situation dependent. When your purpose is to communicate nonverbally with a member of the opposite sex, do not use paper matches or disposable butane lighters. Instead use a high-quality, well-made lighter.

Around the warmth of a campfire, lighting the tobacco from a piece of glowing kindling you choose and pick up is as enjoyable, if not more enjoyable, than smoking the cigarette itself.

Lighter Skills

Zippo and similar lighters with spring hinges and polished steel bodies can be opened, closed, and operated in a number of interesting ways. In 1996, Morten Kjølberg of Norway developed a Web site called www.Zippotricks.com that explored this phenomenon. The Web site provided instructions on unusual ways of manipulating a Zippo lighter. People enjoyed both learning these tricks and seeing others do them.

According to Kjølberg, smokers and nonsmokers alike have been doing sleight-of-hand tricks with Zippo lighters since 1933, when the first Zippo lighter was manufactured. His Web site attracted millions of visitors in short order.

But individuals and organizations poisoned with a nanny-state mentality applied pressure to Kjølberg's backers, who in turn induced Kjølberg to shut down his site. His critics evidently believed that free, literate adults couldn't be trusted with such sensitive information as to how to manipulate cigarette lighters. The more hysterical among them warned of imminent lighter-trick-related carnage and even managed to link lighter tricks to the tragic 2003 fire at the Station nightclub in Rhode Island.

A widespread protest to the perceived interference in an individual's right to freely obtain information induced Kjølberg to reconsider. After receiving 20,000 e-mails of support, he started a new site called www.lightertricks.com, containing a multitude of lighter moves and routines. The Web site is very popular.

Any cigarette lighter trick routine is merely a combination of basic manipulations—the building blocks. While the manipulations or tricks may sound simple, most require considerable practice to master. Based on the number of lighter trick videos available on the Internet, the number of people with well-developed lighter skills is quickly increasing. Visit www.youtube.com, www.metacafe.com, or www.lightertricks.com for hundreds of videos featuring skilled lighter tricksters.

Before undertaking lighter skills, take a moment to consider safe working practices. Lighters produce fire and are dangerous if improperly handled. Practice manipulations in areas where accidental drops will not result in anything catching fire. My personal experience indicates that learning to perform lighter tricks and routines includes many drops, so softer, resilient surfaces are easier on the lighter. Practice outdoors but not on dry grass or other flammable surfaces.

To open the lighter with a flourish, hold it with your index and middle fingers on the lid and thumb on the case, hinge pointing away from you. Squeeze the lighter quickly.

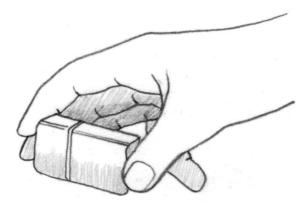

Zippo lighter trick step 1

After the squeeze, your index and middle fingers will be positioned on the backside of the lighter and your thumb on the front side. If you squeeze correctly and with just the right amount of pressure, the lid will pop open.

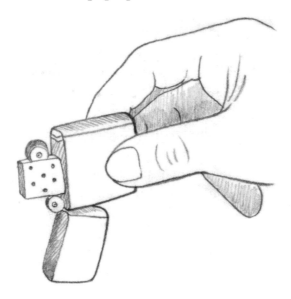

Zippo lighter trick step 2

Morten Kjølberg's Top Tips for Lighter Tricks

• The Squeeze: In your dominant hand, with your palm up, hold the lighter with index and middle fingers on the lid and thumb on the case, with the hinge towards your pinkie. Squeeze the lighter so that your index and middle fingers slide down the reverse case, and your thumb falls on the

front of the case. If you do this quickly enough, the lighter
will open itself.
- The Snap/Flip: This means snapping your fingers in order
  to perform a function. If you snap with your middle finger
  and thumb, then, as your middle finger comes off your
  thumb, it can easily do several things: open the lid, strike
  the flint wheel, or close the lid.
- The Flick: This is the opposite of the Snap. Place the tip of
  your index or middle finger behind the tip of your thumb
  and "flick," causing your index or middle finger to shoot
  straight out—similar to how you would flick a speck of dirt
  off the top of a table. This basic move can be used to open
  or shut the lid or light the flint wheel.
- The Lid Swing: Open or close the lighter's lid by holding
  the case in any manner, then flicking the wrist, causing the
  lid to swing open or shut.
- The Lid Smack: Open or close the lighter's lid by smacking
  it against a surface or smacking something against the lid.

Once you have mastered the basic moves, you can come
up with longer and more sophisticated routines by concate-
nating the basics and adding your own flourishes.

Extinguishing Your Smoke

When you're done smoking, dispose of the butt by stubbing it out
in an ashtray or throwing it into the embers of the campfire.
Never throw the stub on the ground. Besides being dangerous,
that communicates a lack of class.

Absinthe

Absinthe was the most popular drink in 19th-century France. The icon of the bohemian life, *l'heure verte,* "the green hour," was a daily event among hip European imbibers. Indeed, the image that often comes to mind when considering absinthe is a street full of dissipated Parisian intellectuals, some of whom have sunk into poverty and madness by dancing a bit too closely with the Green Fairy.

The most well-known absintheur is Vincent van Gogh. Long unknown and impoverished, he became famous and successful only posthumously. Van Gogh was a clinically depressed epileptic and a social outcast who drank a whole lot of absinthe. He famously shared rooms with Paul Gauguin in Provence for several weeks—until he sliced off his ear in a fit of rage. In 1889 the townspeople of Arles forcibly sent him to a mental hospital to rid themselves of their frightening, alcoholic neighbor.

Was van Gogh plunged into madness by absinthe, or was it merely the outlet chosen by an already disturbed mind? More to our point, was absinthe truly more dangerous than other alcoholic drinks of the day?

It probably was, but not because of any hallucinogenic or narcotic chemical contained in the reputedly dangerous wormwood from which absinthe is distilled. Some researchers say it was the drink's extremely high alcohol content, required to keep the natural oils in suspension, that made it dangerous. Others claim it was the way the drink was manufactured, not the ingredients, that earned the stuff its bad reputation. According to *Scientific American,* many grades of absinthe were produced in late 19th- to early 20th-century Europe. Low-cost, low-grade absinthe, which accounted for the majority consumed, was true rotgut, often adulterated by cheap, poisonous chemicals such as antimony salts and copper sulfate.

Because of too many van Gogh–like outcomes, the liquor was made illegal just about everywhere. In 1906 Belgium forbade its sale. In rapid succession, so did Switzerland, France, the United States, Australia, Germany, and most other countries. After 1920, the liquor continued to be distilled in only a few places, notably Spain and Bohemia.

Prohibiting absinthe production led to the special sort of allure often associated with the officially forbidden. Distilled from wormwood, the bitter-tasting stuff was hard to obtain and had a reputation for providing a more powerful and dangerous chemical kick than other forms of alcohol.

Most modern research suggests that the latter part of its reputation is undeserved. However, in the world of the minor vices, image is often more important than fact. Absinthe, more through its history than through its chemistry, enjoys a rarified position among imbibers.

Dr. Dirk Lachenmeier is head of the Alcohol Laboratories at the Chemical and Veterinary Investigation Laboratory in Karlsruhe, Germany, known by its German acronym CVUA. The CVUA is one of the most renowned institutes in the world in terms of analyzing and evaluating alcoholic beverages. A government-financed organization, it takes no money from the alcohol industry and is not associated in any way with absinthe producers.

Lachenmeier's research into the nature of the psychoactive chemicals in absinthe has done much to clear the misconceptions and fallacies that little-t worrywarts used to ban and overregulate absinthe.

Dr. Lachenmeier provided me with some interesting insights. With his assistance, I've developed a strategy for connoisseurs residing in the Golden Third, anxious to enjoy the novelty as well as the historically significant experience associated with enjoying *la fée verte*, the Green Fairy. When you partake of the drink, you follow in the footsteps of fabled edgeworkers van Gogh, Hemingway, Picasso, and Wilde.

In the 1990s the European Union instituted a single standard to replace the hodgepodge of country-by-country rules regarding the production of absinthe. Joyfully, countries with long histories of absinthe prohibition and even longer histories of absinthe drinking are once again producing liquors similar, if not identical, in quality and composition to the products of the 1890s.

However, there is wide variation in the quality of products labeled as absinthe, ranging from excellent to miserable. Some manufacturers merely mash together a bunch of chemical extracts, add green vegetable dye and ethanol, and proclaim it to be absinthe. Whether such products should truly be considered absinthe is doubtful. Complicating matters is the confusion surrounding the chemical constituents of the liquor, most notably the chemical called thujone, which is allegedly the psychoactive (and therefore restricted) ingredient in wormwood. It can be difficult to determine the best way to actually experience absinthe.

I asked Dr. Lachenmeier to provide simple guidelines for obtaining and enjoying authentic (and legal) absinthe. The first is to partake responsibly and in moderation.

Dr. Lachenmeier's Top Tips for Selecting Absinthe

· Absinthe should be distilled, not mixed.

Bottles labeled as the alcoholic beverage absinthe are produced in one of two ways: distillation or mixing. Absinthe produced by distillation is far superior and is really the only acceptable choice for an authentic and enjoyable experience.

Quality absinthe production begins by steeping worm-wood and other dried herbs such as fennel and anise in ethanol. After many hours, the macerate becomes greenish and is very bitter and strong smelling. Water is then added, and the product is distilled. After distillation, additional wormwood is added along with other herbs such as hyssop and lemon balm. Finally, the mixture is diluted with water until the desired drinking strength is attained.

· Absinthe should be 90 proof or higher.

Absinthe is by nature a strong drink. It should have an alcohol content of at least 45 percent, and alcohol levels of 75 percent are not unknown. Lesser proofs will not dissolve and properly suspend absinthe's essential oils.

· Quality absinthe is naturally green or, occasionally, colorless.

The characteristic green color of absinthe should be achieved only with wormwood and other herbs. The introduction of artificial food dyes or colorants (check the ingredients on the label) denotes inferior production.

· The bitter flavor of wormwood is the main taste characteristic of absinthe.

If it doesn't taste like wormwood, it's just not absinthe. There's no such thing as mint-flavored absinthe.

· It clouds when diluted with water.

When water is added to absinthe it becomes milky green. The opaline coloring is due to the sudden insolubility of the resinous oils that occurs when the concentration of alcohol within the glass is reduced. This is called the louche effect (louche is French for disreputable or indecent) and is a traditional, important part of the absinthe drinking experience.

Once you procure some quality absinthe, how do you go about enjoying it?

In the tried-and-true Gallic ritual, the absinthe drinker pours a one-ounce portion or "dose" of absinthe into a glass. Next, a special perforated spoon is balanced on the rim of the glass and a sugar cube is placed on the spoon.

A thin stream of very cold water is carefully dripped on the sugar cube, which dissolves slowly into the absinthe below. When the cold sugar water mixes with the absinthe, the result is the *louche*, the opaque, opalescent color change heralding the arrival of the Green Fairy.

Interestingly, the chemical processes are the same ones we exploited to make gunpowder via the Frankfort Arsenal's field expedient method in chapter 7. As we saw before, alcohol reduces water's ability to hold chemicals in solution. This time, instead of precipitating out gunpowder, the essential oils in the liquor's botanical extracts come out of the solution as the milky green-white *louche* so loved by connoisseurs.

In the classic absinthe ritual, the drinker takes time when dripping the water on the sugar cube, a single drop at a time. The imbiber often stares into the glass watching the *louche* grow and swirl with each additional droplet of sugar water.

How much water should be added to the absinthe? It depends on personal taste. To begin, start with a ratio of four parts water to one part absinthe and adjust up or down to suit your prefer-ence. *À votre santé*!

Driving Fast

You can barely see at a hundred; the tears blow back so fast that they vaporize before they get to your ears. The only sounds are the wind and a dull roar floating back from the mufflers . . . howling through a turn to the right, then to the left, . . . letting off now, watching for cops, but only until the next dark stretch and another few seconds on the edge.

—HUNTER S. THOMPSON,
HELL'S ANGELS, 1966

Speed thrills. Nothing on Earth provides the same type of thrill as pedal-fully-down, straight-line acceleration in a fast car. If rocketing from 0 to 60 mph in under 7 seconds is exciting, then doing it in under 4.5 is indescribable.

Going faster means going better. It's the feel of your back pressing hard against the firm resilience of the seat cushions; new vistas careening into view and then out of sight, trees and billboards a receding blur in the rearview mirror. It's speed-shifting through five gears, your right hand a blur as it moves between the steering wheel and shift knob.

Speeding is a vice and a thrill. Of course you're aware that exceeding the posted limit is illegal, that speeding kills people, that the police will throw you in jail, that your car's mileage will be reduced to that of a ferryboat in a headwind. Acceleration like that isn't for anytime and anyplace—or for everyone. That's why, for better or for worse, we have speed limits. Speed limits are often controversial. Set too high or too low, they engender complaints and anger. In the 1970s, the federal government enacted legislation designed to conserve gasoline and diesel fuel. This resulted in the hated double nickel—55 miles per hour—speed limit. Traveling 55 on straight, modern freeways in a well-maintained car seems odd and unnatural. It's like watching a Jamaican sprinter trot.

The reduced speed limit was a response to the 1973 oil crisis. The U.S. Congress and President Nixon imposed the nationwide 55 mph speed limit in 1974, enforced by the threat of withheld highway funding to any state that refused to enforce the law. While some analyses indicated the lower speed limit was effective in terms of energy conservation in urban areas, other studies showed that the reduced speed limit was far less effective at reducing energy consumption than its proponents claimed.

Advocates of the 55 mph limit also claimed that the lower highway speeds would improve safety. But the studies undertaken to prove the claim were inconclusive at best. In fact, according to the *Wall Street Journal*, the lower speed limits eventually worsened, not improved, safety on the interstates. Despite its unpopularity, the 55 mph speed limit was retained after the oil crisis abated, a nanny-state measure designed to protect the populace from a danger more imagined than provable.

As popular as herpes, the 55 mph limit was the most frequently broken law since prohibition. In the mid-1980s the law was repealed, and speed limits were increased to 65 mph in most states. Over time, legal speeds have risen to 70 or even 75 mph, but that hardly approaches the velocity needed to provide a thrill and make the heart race.

There will never be an American equivalent of the autobahn, the German roadways where speeds are limited only by the ability of a vehicle to couple horsepower to a drive shaft and the ability of the driver to guide it safely.

That's probably a good thing.

The Physics of Speed

Kinetic energy is an interesting quantity, the mathematical function that describes the result of bringing a moving body to complete and sudden stop. It is a second-order function, meaning it increases proportionally to the square of something, in this case, the body's speed. A vehicle traveling at 20 mph has four times the energy of one traveling at 10 mph. A vehicle traveling at 40 mph has eight times as much.

In situations where a vehicle crashes head on into an unmovable object, higher speeds exponentially increase the violence of a crash. And studies show that the probability of fatality in a collision rises even faster; the likelihood of highway death accelerates on a sky-rocket-like fourth-power function proportional to a vehicle's speed.

But these sorts of forces come into play only in head-on collisions or direct hits on immovable objects such as concrete bridge abutments and trees. Although there is a clear relationship between vehicle speed and the severity of a crash, there are few instances of such collisions on high-speed, interstate highways. The design of roadways reduces the possibility for most head-on conflicts. But the interstate highway is a far from ideal place to determine the maximum attainable speed of an automobile or motorcycle.

Where to Drive Fast

The first speed limit was the 10 mph limit introduced by the Locomotive Act of 1861 in the United Kingdom.

In the United States, expressway and interstate highway speed limits vary from state to state, but are capped at 75 mph, with one small exception in parts of rural Texas where 80 mph is allowed.

STATES WITH AN 80 MPH SPEED LIMIT	
Texas (on parts of I-10 and I-20 in west Texas)	
STATES WITH A 75 MPH SPEED LIMIT	
Arizona	Nebraska
Colorado	Nevada
Idaho	New Mexico
Montana (When the federally man-	North Dakota
dated limits were first repealed,	Oklahoma
Montana briefly played with a	South Dakota
speed limit of "reasonable and	Utah
prudent." The "Montanabahn"	Wyoming
lasted only a few years and now the	
trip from Lewiston to Great Falls	
is controlled by speed limits.)	

A bit surprisingly, the countries with the highest speed limits
are in the Eastern Hemisphere. Probably the highest speed limit
is 160 km/h (100 mph), posted on the A-10 Tauern Motorway in
Austria and the Emirates Road in Dubai, United Arab Emirates.
At that speed, the lines on the highway look like dots, and the
scenery speeds by like a DVD on fast forward. But still, you may
ask, are there places where you can go faster? Well, a few public
roads have no speed limit at all.

For example, the German autobahns are famous for having no
universal speed limit, although about a third of them do have
posted speed limits, and limited sections are equipped with
motorway control systems that display conditional speed limits.

On the open autobahn, it is not uncommon for BMWs and
Porsches to rocket by in the passing lane at speeds over 125 mph,
but most drivers travel at much lower speeds, usually around 80
to 95 mph. Obviously, speeds are naturally limited by traffic den-
sity and the abilities of individual cars. Those with smaller
engines do not have the horsepower to exceed 100 mph, and the
manufacturers of most fast luxury cars install electronic governors
that limit the maximum speed of their products to about 155
mph. So, it is nearly impossible to drive at high speeds for more
than a few miles at a time.

Certain jurisdictions in India and Nepal simply don't have speed
limits on the law books. But given the crowded conditions and poor
quality of highways, maybe they don't need them. Regardless, they
seem like poor choices for road testing your new Jaguar XJ.

Most of the United Kingdom has a maximum limit of 113
km/hr or 70 mph on its expressways. But unlike the rest of the
British Isles, there is no national speed limit on the Isle of Man.
There are speed restrictions in built-up areas, but after you pass
the National Speed Limit sign, there ceases to be any speed
restriction. Nonetheless, people driving in a reckless manner or
talking on cell phones while driving will likely be stopped by
authorities.

This tiny island is home to the Tourist Trophy motorcycle race and has long had a reputation as a gearhead Mecca. Because of its close association with the high profile, high-speed motorcycle race, many Manxmen call their abode the road racing capital of the world. The event draws more than 40,000 spectators each May. For many high-performance cyclists, accelerating their bikes legally to speeds exceeding 100 mph along the 15-mile-long Mountain Road between Ramsay and Douglas is their idea of cycling heaven.

The high-speed lifestyle of Isle of Man residents comes with some pretty serious baggage: more than 100 Manxmen and women have been killed and 10 times that many seriously injured since 1993. Speed was at least a partial cause in more than half of these accidents. Even so, the idea of a speed limit is anathema to Manxmen; multiple polls and surveys show that the overwhelming majority of residents oppose speed limits.

Open Road Racing

Like tobacco, alcohol, and promiscuous sex, driving fast may not be healthy or smart, but for some the urge to live dangerously exists deep within the genes. Some people are going to do it anyway. Driving very fast is dangerous to others as well as to the direct participants, and can't be condoned on most roadways.

Since most erstwhile road rocketeers don't have the wherewithal to ship their prized Vettes and Shelbys to Frankfurt for an autobahn cruise, an alternative called Open Road Racing may be one answer.

Open road races are held several times a year on selected highways in Nevada and Texas. The race organizers obtain permission from local and state authorities to close the roads to normal vehicular traffic. Once the road is clear, racers leave the starting line one at a time, at roughly one-minute intervals. The fastest cars start first to lessen the chance of any overtaking or passing. Cars compete against the clock, the objective being to match the average speed over the course with the targeted speed.

Most types of vehicles are permitted, from normal automobiles to purebred racing machines. The maximum speed allowed a particular car and driver is determined by the vehicle's safety equipment and the driver's experience level. The racecourse's length is typically 50 to 100 miles long, although there may be multiple passes through sections.

How fast can you drive? It depends—mostly on how much money you have. Just like an automobile's kinetic energy, the cost of automobiles rises exponentially with top speed. Speed classes start at 80 mph and go up. Top speed classes in some open road races push 200 mph.

How to Drive Fast

Gale Banks may be the world's best-known gearhead. He's the president of 200-employee Banks Power, Inc. His company makes automotive speed equipment and manufactures specially designed or "tuned" vehicles: modified cars and trucks capable of hitting speeds in excess of 200 mph. He's also the man Jay Leno calls when he has a question on tuning one of his many muscle cars.

Banks has been building and driving fast cars since he goosed his mom's 1931 Model A Ford to make 100 mph. In the 50 years of experience he's accumulated since, his vehicles have gone only faster and faster.

Banks has several projects of which he's particularly proud. There's the 240 mph Pontiac Firebird he built back in the 1980s. His current project, a Dodge Dakota named Project Sidewinder, is the world's fastest diesel truck, capable of speeds exceeding 220 mph (and it gets 24 miles per gallon when driven at normal highway speeds). The fastest of all is the Streamliner, a specially designed land speed racer that top-ended at an incredible 432 mph on the Bonneville Salt Flats. Now that's living dangerously.

Gale Banks's Top Tips for How to Drive Fast

- Know thy vehicle.

Prior to going fast, make sure your ride can handle the speed. There are several things to consider prior to stomping down on the accelerator:

First, check your tires for inflation and construction. Tires are rated for a maximum speed and no more. Exceed the rated speed and there's a chance the plies could come loose, leaving far more rubber on the road than you want.

Second, make sure your shocks and other suspension equipment are in good working order. Shocks are required to damp vibrations in the suspension. At high speeds, worn shocks won't hold up to bumps, making control of a vehicle difficult if not impossible.

Third, your brakes must be in good order and capable of stopping a fast-moving vehicle if an emergency should occur.

- Know thy area.

Areas with moderate to high traffic levels are poorly suited for fast driving. Also, some lesser-traveled areas are known for particularly rigid or loose enforcement of traffic laws. It may be advisable, then, to choose areas for driving fast accordingly.

- Know thy road.

Before driving fast, drive slow, real slow. It's important to inspect the course prior to making a speed run, making sure there are no obstacles, debris, or loose gravel that could pose a problem at higher than normal driving speeds. Take note of soft shoulders, turns, or curves in the route.

- Know thyself.

At higher speeds, your reflexes and knowledge determine your ability to respond to situations. Don't exceed your capabilities or your vision.

Persiflage

Persiflage means to treat a subject or topic flippantly. Close approximations of the term include *hoaxes, jests,* and, in a manner of speaking, *lying*. A practical joke is an example of persiflage. But persiflage is bigger than that, for it incorporates an element of risk, wagering a fine or a punch in the nose, against a chance to make a really good joke.

As I mentioned in the opening chapter, I spent a lot of time, too much time, in occupations well suited for someone on the far left side of the risk curve. Good jobs for Caspar Milquetoast or J. Alfred Prufrock perhaps, but not for me.

When I first worked at the phone company, I was given a parking spot in the company lot next to the building. My job required frequent daily trips to and from the office. Then one day my world, such as it was, was shaken. The company announced that henceforth, those parking spots would be reassigned to upper level managers, a perquisite of their power and status. I was annoyed.

The phone company had its special way of doing things, a complex bureaucracy so Byzantine it would make a French civil servant cry, "*Ça suffit!*" There were forms for everything, and those forms would be couriered between cities and locations in brown envelopes tied with string.

Phone company forms were the veins and arteries of the entire business, flowing into and out of each departmental organ, nourishing each with the bureaucratic and accounting authority needed to survive.

The night after the parking change, I waited until after six o'clock. Alone in the building I sat down at the single personal computer in our office, a Macintosh Plus with a rudimentary page layout program. There I created "Phone Company Form PF 11421—Request for Parking Space Allocation."

PF 11421 was a five-page, double-sided application requiring those who had been assigned parking spaces in the recent reallocation to provide details about themselves and their auto "for company insurance and record keeping purposes."

Instructions for Completing PF 11421 (excerpt)

- *Complete the attached three-page health questionnaire (Sample question: Do you have any bladder control or enuresis (bedwetting) issues affecting your ability to drive or maintain your vehicle's appearance?).*
- *Attach a 4" x 6" color photograph of your license plate.*
- *Provide auto insurance policy numbers and contact information for all policies you've had for the last 12 years.*
- *Provide your car's VIN (vehicle identification number) and engine block identification number.*

I'm not sure there is such a thing as an engine block identification number, but several form recipients steam cleaned their engines to look for it.

The form, emblazoned with a rather snappy logo from the mythical Phone Company Parking Authority, was distributed to those once-happy managers in the official string-tied intracompany mail envelope.

Most recipients grumbled, then bit the bullet and started to gather the information. Some managers grew angry and actually started shouting. A few—the most frumpish company men—loudly defended the company's need for the form. In the end, nearly every manager dutifully filled out PF 11421 and sent it to headquarters, where it was no doubt routed, stamped, signed, routed again, and stored.

But I'm just an amateur compared with those who are really talented.

On October 15, 1941, the *New York Herald Tribune* reported that Plainfield State Teacher's College had defeated Winona 27–3 in football.

In the weeks that followed, most New York papers closely followed the fortunes of tiny but powerful Plainfield State, led by Chinese-Hawaiian fullback Johnny "The Chinese Comet" Chung. The Plainfield juggernaut rolled on and on, defeating Chesterton, Appalachian Tech, and St. Joseph's. Plainfield's team was racking up victory upon victory, due in no small part to Chung and coach Ralph Hoblitzel's unique W formation, in which the ends line up backwards, facing the halfbacks. This unique formation benefited the talented Chung by providing better blocking, because as stated in a Plainfield press release, "in this way our ends can see immediately who has the ball."

The '41 Plainfield squad had some of the best stats in college football history. But what was unique about this team was that they existed only in the mind of Morris Newburger. Newburger was a stockbroker with a sense of humor. On a whim, he called the sports desk at the *Herald* with the bogus score. When he saw the results the next day, he was delighted. Each week thereafter, he called in PSTC scores and highlights, going so far as to send out press releases and backgrounders, which led to numerous articles in the major New York dailies. Eventually he was found out, betrayed by a secretary who overhead Newburger plotting Plainfield's final victory (a 40–27 beat down of Harmony Teacher's College) and blew the whistle.

A more recent example of persiflage is the Bonsai Kitten. It's an elaborate Internet-based hoax that features a Web site purporting to provide information on the unusual practice of growing kittens in glass jars. The kittens purportedly acquired the shape of the jar as they grew.

At Bonsai Kitten, we achieve this by placing the kitten into a rigid vessel soon after birth, and allowing the young cat to grow out its formative time entirely within this container. The kitten essentially grows into the shape of the vessel! Once the cat is fully developed, it is removed (or the vessel bro-

ken to remove it), producing the lovable, furry pet you've always wanted, but it remains in the shape you've always dreamed of! There is virtually no limit to the eventual shape of your pet.

—BONSAI KITTEN WEB SITE

(A copy of the original site is at www.shorty.com/bonsaikitten/index.html)

This elaborate hoax was dreamed up by a group of MIT graduate students. The bottled kitten allegedly breathed through specially drilled holes in the glass, through which it also ate and expelled waste. Ludicrous it may be, but an amazing number of people believed that a man named Dr. Michael Wong Chang was indeed promoting the practice of stuffing kittens into small glass jars to reshape them "to your exact specifications."

Was this stunt Big-T type stuff? You decide. The Web site sparked threats, denunciations, and even an FBI investigation. Thousands were outraged. But it's edgy, clever, and to many people, a very good joke indeed.

Perhaps the best-known current prankster is Alan Abel. Abel's prankster career began in 1959 when he formed SINA, the Society for Indecency to Naked Animals. Ostensibly a hyperconservative political organization dedicated to the aim of clothing naked animals, SINA was taken seriously and featured in numerous national television programs including the *Today Show* and the *CBS Evening News* with Walter Cronkite. SINA's slogan was "A nude horse is a rude horse."

Over the years, Abel has pulled off more than a dozen major hoaxes including appearing on national television as Omar, founder of Omar's School for Beggars. This was supposedly a training camp for professional panhandlers. Disguised as Omar, Abel was interviewed (and usually angrily denounced) on national talk shows by Tom Snyder, Morton Downey Jr., Sally Jessy Raphael, and Mike Douglas.

In 1999 he appeared on an HBO program called *Private Dicks, Men Exposed,* a show in which "a wide variety of male specimens boldly go before the camera to discuss their relationship with their penises." Abel told the producers that he was Bruce Spencer, a musician from Ohio, and held the *Guinness Book* record for the world's shortest penis. The show's producers took him at his word and featured him prominently in the show, talking at length about his tiny member. Only after the show aired did they, to their extreme embarrassment, find out that they'd been had.

Abel's motivation is a bit complex. Sure, it's funny, but for Abel, there's more to it. There's a message. Typically, he says, he does what he does to expose hypocrisy and the media's willingness to air pretty much anything as long as it's salacious enough to improve ratings. Most recently Abel assumed the identity of Jim Rogers, leader of a campaign to ban breastfeeding because "it is an incestuous relationship between mother and baby that mani-fests an oral addiction leading youngsters to smoke, drink, and even becoming a homosexual." Rogers granted more than 200 interviews to outraged yet credulous media types until Abel fessed up in 2005.

Alan Abel's Top Tips for Persiflage

- Don't underestimate the credulity of people. It's amazing what people will believe if you say it with a straight face.
- The hoax has to be compelling. That is, it must capture a person's interest right away. (Abel's stunts include the Euthanasia Cruise, the KKK Symphony Orchestra, and a topless string quartet. All of these, if believed, simply could not be ignored by the media or anybody else.)

> • Timing is important. Hoaxes work best when they're not competing with more important (i.e., legitimate) news for attention. The slowest news days are the best days for pulling off a hoax.
> • The thing that separates great hoaxes from mediocre ones is that the best contain a message behind their madness.

There are other minor vices, including swearing and gambling. What separates a minor vice from a major one? Occasionally engaging in a minor vice under well-defined circumstances bumps you gently to the right on the risk-taking curve without hurling you into the outlier zone.

In the next chapter, the focus shifts to a physical, hands-on form of living dangerously. Perhaps such skills would be handy should your attempt at persiflage send circumstances bounding out of control. These are the physical arts. Specifically they involve the delightful act of throwing things: whips, knives, and punches.

11

The Physical Arts

*"Believe me! The secret of reaping the greatest fruitfulness
and the greatest enjoyment from life is to live dangerously!"*
—FRIEDRICH NIETZSCHE,
THE GAY SCIENCE

A study of dangerous living that excludes the theory and application of the use of knives, sticks, canes, whips, and so forth
would be incomplete. Full understanding of living dangerously
inexorably brings you into full body contact with the hard, the
sharp, and the edgy at some point. Well that time has arrived, and
as Nietzsche suggests, it could well be the surprising pathway to
much enjoyment.

Some little-t critics may see this chapter as excessive in manli-
ness; a testosterone-charged set of activities suitable only in a
lumberjack bunkhouse, at a Civil War reenactors's camp, or on a
pirate ship. I disagree.

These projects are fun and entertaining to learn and practice.
Men and women can enjoy them equally. And they are culturally
significant, as they are well rooted in science and are frequently

encountered in history and literature. But nowhere do they show up more prominently than in movies.

Bullwhips

The 1980 movie classic *The Blues Brothers* has a scene I really love. Joliet Jake Blues (played by John Belushi) and his band are onstage at a roadside dive called Country Bob's Bunker. The tough crowd doesn't seem to be enjoying the band's music, and the chicken-wire screen in front of the band is doing only a mediocre job of protecting the band from a fusillade of domestic beer bottles.

Looking off to the side, Jake spies a mean-looking bullwhip hanging on the wall. A lightbulb flashes on in his head. He turns to his band and asks, "You guys know 'Rawhide'?"

> *Move 'em on, head 'em up*
> *Head 'em up, move 'em on*
> *Move 'em on, head 'em up*
> *Rawhide! (WHIP CRACK)*

With each chorus, with each crack of the whip, the crowd's demeanor changes. By the time the Blues Brothers finish the song, the crowd is won. "Rawhide" is a good song, but it's not the lyrics that do the job. It's the whip. Everyone, from cowboys to blues singers, loves the sound of a whip crack.

Whips are interesting on a variety of levels. Music aside, whips were traditionally used as herding and cattle-driving tools. A good handler can throw the whip in such a manner that the loop coursing through the bullwhip's length reaches incredible speed. By the time the roll approaches the end, the loop velocity has gone transonic, crashing through the sound barrier: *Crack!* Back on the trail, cattle drivers used this noise—in actuality a sonic boom—to control herds of Herefords and Longhorns.

A whip is no mere rope on a handle. Rather, it's a complex system of finely tuned interacting parts. Crafting a whip that cracks well requires time-tested, specific methods. In several important ways, a whip differs from other long, flexible members such as cables and thongs. A whip must taper gradually from base to end in a specific fashion, and it must be made of materials that gracefully and easily transfer the energy and momentum of the moving internal coil along its length. A good whip is engineered such that the momentum and energy are smoothly transferred from the thick plaits near the handle to the thin, flexible popper at the end. A well-designed whip is a lesson in supersonic fluid mechanics.

Alain Goriely, professor of mathematics at the University of Arizona, is intrigued by the phenomenon of whip cracking. He has published several research papers on the topic and has come up with fascinating conclusions about the way waves propagate through flexible devices such as whips. He told me that his research shows that the whip-throwing motion causes a loop to propagate down the length of the whip, its speed increasing in a mathematical progression inversely related to the thickness and weight of the whip's cross-section. Dr. Goriely developed a series of equations that accurately model the curvature, boundary conditions, tension, and speed of the whip. He then fed that data into a computer. The results of this study in whip dynamics show that the cracking sound results from the hypersonic movement of the loop or coil through the fibers in the thong of the whip. Goriely has measured tip speeds exceeding Mach 3 in a well-made whip. So your basic bullwhip is nothing short of a human-powered, mini-sonic-boom effector.

By its nature, the motion involved in whip throwing makes it a dangerous proposition. The tip speed of a bullwhip under the control of an expert can exceed the sound barrier. A leather thong flying through the air at great velocity in close proximity to humans inherently involves some peril.

The first time I tried whip cracking, I found the activity hard to master and frustrating. Yet there was something inexpressibly interesting in the activity, in attempting to master the eye-arm-body coordination required to handle a plaited leather whip. I persevered and am proud to say I am now able to crack a whip with fair precision. If you try, you can too.

All About Whips

A whip has four parts: handle, thong, fall, and cracker. The largest is the thong, a section of leather braid wrapped around a flexible core called the belly. The thong attaches to a handle, which is usually a short section of wood, sometimes bare and sometimes covered in leather or cord. The handle is shaped to rest comfortably in the hand.

At the end of the thong is the fall, a series of specialized leather knots connected to a thin, replaceable tail. A foot or two long, the fall is typically a single piece of leather a bit wider than the individual braids at the end of the thong.

Finally, a small piece of string or nylon cord is attached to the fall at the very end of the whip. This is the called the *popper* or *cracker*. The cracker moves fast, very fast, and is subject to a lot of wear. Frequently used whips go through a lot of crackers.

Professor Goriely divides the world of whips into two categories: pain-making whips and noise-making whips. The pain-makers, such as the infamous cat-o-nine-tails, are short, bulky, and made up of several strands that hang separately. They don't crack. Noise-making whips are long and tapered, consisting of a single braided strand. Despite their frightening appearance, noise-making whips are not used as weapons or torture devices. Their purpose is to produce incredibly high tip velocities, causing very distinct, loud cracks.

Whips must be well made in order to crack consistently. Even an expert will have difficulty cracking a poorly made whip with loose plaiting and frayed edges. The best-quality whips are very

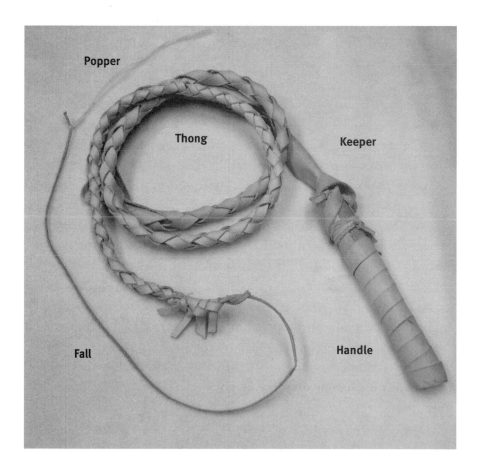

expensive. However, whips of reasonable quality can be purchased from reputable whip makers at fair prices. The Australians are considered very good whip makers and the best of them use braiding made of kangaroo leather. Less expensive whips are made of cowhide.

Cracking Your Whip

Learning to crack a whip may be harder than you might expect. Not only that, until you get a feel for how to handle a whip, you will probably get a few welts. But most thrill seekers would take that trade-off in a heartbeat.

Remember:

Eye protection is absolutely required. A hat that covers your ears is a very smart precaution.

Wearing gloves, a jacket, boots, and heavy pants will reduce your chances of getting a whiplash.

Keep people, animals, and things you don't want broken clear of your practice area.

My foray into the world of whips led me to a man near its center: Robert Dante, an affable, silver-haired bullwhip expert. Robert is a professional whip artist and a leading authority regarding the use of whips. In fact, he was the man who broke the 200 cracks per minute barrier, cracking his six-foot whip an incredible 203 times in 60 seconds at an official trial in 2004. That's more beats per minute than a drum corps flushed with espresso, and good enough to establish the Guinness World Record for most bullwhip cracks per minute, although he's since been overtaken.

According to Dante, there are many methods of whip cracking. All of them require practice to obtain consistent cracking. Practice makes perfect (and builds arm strength)!

Robert Dante's Top Tips for Whip Handling

The basic method of whip cracking is called the circus crack. Here is Robert Dante's advice for learning the technique.

- Put on safety glasses and a hat.
- Swing the whip up vertically into the air over your head. Your hand should end up at about two o'clock in front of you. You should feel the tug of the cracker. Now the whip is traveling backward, over your shoulder, still on a vertical plane.
- At this point you bring your elbow into play. When the handle is pointing to two o'clock, break your elbow and drop the upper arm slightly, and then push forward. This makes a little circle in the air with your fist, similar to the circle made by the drive wheel of a train's locomotive. Making this form in the air makes the whip create an *S* shape and start rolling forward, slightly under the path in the air it just made coming up over your shoulder.
- As you move your hand forward, the top of the S rolls out, ending with a crack.
- Once you have mastered this, then squeeze the handle, as if you're pulling a trigger. Keep the whip pointed at the target. The whip will go into a "supercharge" mode, and the crack will be louder, with no loss of crispness. When you hear the crack, the energy has been expended.
- Much of the technique of whip cracking comes down to controlling the speed and location of the whip loop during the throwing action. You don't need to throw hard; let the whip do the work.
- The quality of your whip makes a tremendous difference. The better the quality of the whip, the better it will handle.

Knife Throwing

Alonzo is an apparently armless knife thrower who uses his feet to encircle Estrellita with blades. Estrellita falls in love with Alonzo (she fears men's arms), so he goes to a hospital and has his amputated. Meantime Malabar cures Estrellita of her fear of men's arms, so Alonzo tries to have him killed during a circus act.

—ED STEPHAN'S PLOT SUMMARY
FOR THE UNKNOWN, 1927,
METRO GOLDWYN MAYER

Above is the synopsis for one of the strangest movie plots of all time. Fittingly, it involves knife throwing. Knife throwers, as portrayed in the movies, are usually strange and menacing, from crazy Bill the Butcher in *Gangs of New York* to the murderous twins Mischka and Grischka in Big-T archetype Ian Fleming's *Octopussy*. Why the strange karmic cloud that hovers over knife throwing? Mostly, I reason, because of the easily imagined possibility of danger that is inextricable from the act. Certainly, throwing a knife feels like a dangerous action.

But when I do it, I don't find it strange or sinister. Instead, I find it exhilarating. A bit dark perhaps, but thrilling nonetheless. I like the feel of a black, carbon steel, 12-ounce throwing knife in my hand. I enjoy squaring up to the wooden target, taking a breath, extending my arm sharply, and letting the knife fly. Off it goes, turning exactly one and one-half times on its way to the red circle in the center, then with a solid *thwack* sticks there. I will always remember the first time that happened.

It's a much different experience than, say, throwing pub darts. You really can't compare the bold, red-blooded flush of satisfaction derived from a perfect, 90-degree knife stick in a target with the rather dainty, epicene feeling one gets when tossing a dart.

I have spent hours practicing knife throwing techniques, and I can't say I've mastered it yet. Too many of my throws still send the knife banging into the target and falling to the ground. But those throws that do stick in the target more than make up for the ones that don't. The first time all my knives bit into wood and stayed put, I couldn't help but give a shout of satisfaction.

Knife throwing, outside of Hollywood, is a colorful and historically significant activity that conjures up images of Kentucky backwoodsmen or a Buffalo Bill Wild West Show. While satisfying, throwing anything with a sharp edge is inherently dangerous. But if you take reasonable precautions you can mitigate most of the risk.

While knives were important tools going back to prehistoric times, the history of the precisely weighted throwing knife is much more recent. Knives, like other tools, are designed for specific purposes. Heavy kitchen knives are great for chopping, daggers are perfect for piercing, and large-bladed machetes and bolos serve a number of purposes for outdoor enthusiasts and adventurers. But none of these designs is particularly suited to throwing.

The first knife to be really useful as a throwing knife was the famous Bowie knife. The Bowie, reputedly invented by Jim Bowie and his brother Rezin, is a long-bladed, all-purpose knife with a solid hilt and handle guard. It was soon determined that it threw pretty well if you took some time to practice. Civil War photos show troops throwing Bowie knives at targets for sport.

While Union and Confederate soldiers hurled their knives to pass time during long stretches of camp life, it wasn't until the 1950s that modern throwing knives were developed. The modern throwing knife differs considerably from most other knives. Its particular length, weight, and balance allow it to be thrown consistently and accurately. A throwing knife is typically between 12 and 15 inches long and weighs between 10 and 16 ounces. Shorter, lighter knives are difficult to grip and to throw consistently, while longer, heavier knives require more muscle power than most people find comfortable.

Throwing knives

Materials and Tools

- (3 or more) Throwing knives, 10 to 16 ounces each. Many sources of throwing knives are available online. Use a search engine to search for "throwing knives."

- (3) 2-by-12-inch pine boards 3 feet long and cut into an approximate 3-foot circle (or a square if a circle cannot be cut)
- (1) Sheet of plywood for backing

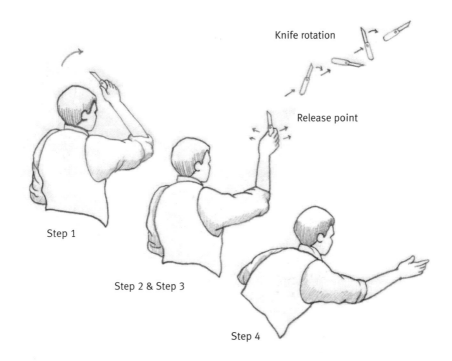

Knife rotation

Release point

Step 1

Step 2 & Step 3

Step 4

As a skill, knife throwing is not particularly difficult to learn, although attaining any real degree of competence requires a fair amount of practice. As the late Harry K. McEvoy wrote in his classic book on the subject, *Knife Throwing: A Practical Guide*, if knife-throwing newcomers have enough coordination to throw a baseball or crack a whip, they can learn to throw a knife correctly.

Like dart throwing, the goal in knife throwing is to hit a target. Unlike dart throwing, however, knives do not travel through the air point first, maintaining an arrow-like trajectory. Instead, a knife rotates, revolving around its center of gravity, as it heads towards the target.

Step 1. Raise your arm over your head with your elbow bent slightly.

Step 2. Swing your arm, from the shoulder, forward and downward.

Step 3. Release at the point shown in the illustration. Open your hand to release the knife without snapping your wrist.

Step 4. After the release, follow through, swinging your arm toward the target.

If the sharp tip is facing forward at the moment of impact, the knife sticks into the target with a satisfying twang. If the handle hits the target, the knife bounces back and falls to the ground with a very unsatisfying thud. The skill required is to figure out how to make sure the knife always hits the target tip first.

The trick is to be very, very consistent. The knife thrower must release the knife in exactly the same way and at exactly the same position each and every time. The thrower determines the distance between the release point and the target face that results in a stick, and throws from that distance each time.

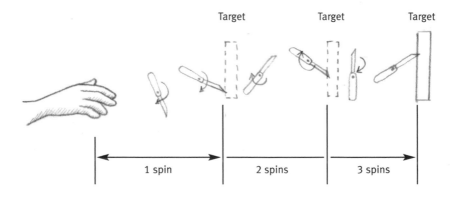

Knife spin

The distance from the release point to the target is a whole number of knife spins. The thrower must figure out how far his or her knife travels in one spin and place the target appropriately. Once consistency at one distance is attained, the thrower can move forward or backward a discrete amount based on how far the knife travels during each revolution.

Knife Throwing Safety

Here are general guidelines for using knives safely.

- Throw only in areas suitable for knife throwing. See the information on targets for rules on finding a suitable location.
- Use knives specially made for throwing. Typically the blade is dull and only the tip designed to stick in the target is sharp. Keep the tip sharp.
- Don't run with knives.
- Don't point your knife at other people.
- Keep the blade of your knife covered in a sheath or store the knife in a box when not in use.
- Inspect the knife frequently to make sure it is in good shape. Make repairs promptly.
- If you injure yourself, seek first aid immediately.

Knife Targets

Placing your target in a suitable area is very important. Find an open area where you can see at least 30 feet in all directions, or set your target against a long continuous wall with no doors and windows. The area should be secured to keep people or pets from accidentally entering the throwing area. It is essential that no person or animal enters the throwing area without the thrower's knowledge.

Throw on grass. Since missing the target will cause your knife to hit the ground, a hard surface will be hard on your knife. On occasion the knife will bury itself in the dirt, making it difficult to find. Keep the grass in front of your target area short to facilitate finding your knives.

The wood grain in the target should run vertically. This permits easier sticking.

Don't use trees as targets. Repeated knife wounds can kill a tree. Instead, use a soft wood (pine is best) target mounted on a large 4 by 8-foot sheet of plywood. Mark the target with five concentric circles each 5 inches wide. The circles can be drawn on wood planks laid side by side, each plank being ¾-inch or more thick. Many experienced throwers leave their targets out in the rain, believing the soaking softens the wood and provides a better "stick."

Knife throwing competitions are similar in nature to archery contests. The most common contest involves assigning a point value to each circular band on the target. Each thrower is given a prescribed number of throws, say 25, and the total number of points awarded is simply the number of sticks in each band multiplied by the band's point value. For extra challenge, throwers typically compete from multiple distances.

Bartitsu

[Moriarity and I] tottered together upon the brink of the fall. I have some knowledge, however, of Bartitsu, or the Japanese system of wrestling, which has more than once been very useful to me. I slipped through his grip, and he with a horrible scream kicked madly for a few seconds and

*clawed the air with both his hands. But for all his efforts he could not get
his balance, and over he went.*

—SHERLOCK HOLMES IN "THE ADVENTURE
OF THE EMPTY HOUSE"

Sherlock Holmes was Sir Arthur Conan Doyle's fictional
detective and an edgeworker without peer. Perhaps Britain's most
popular literary character of the late 19th century, Holmes was
well known for his towering intellect and need for constant mental stimulation. To satisfy his intellectual needs, he engaged in a
number of Big-T activities, including sword fighting, boxing,
and stick fighting, as well as frequent recreational narcotic use.

While Holmes was better known for his reasoning ability than
for his fighting skills, he was quite capable of defending himself
when the chips were down. And to do this, the detective mastered
a now little known but very effective fusion of British boxing
techniques and Japanese martial arts called Bartitsu.

Bartitsu was invented by a British engineer named Edward
Barton-Wright, who combined the martial arts skills he learned
while building railways in Japan with the stick-and-sword fighting
skills he mastered in Europe. Bartitsu drew heavily from stick-
fighting techniques developed by Frenchman Pierre Vigny, English boxing, and Japanese jujitsu. (Barton-Wright coined the
neologism "Bartitsu" by combining Barton-Wright and jujitsu as
a portmanteau word.) Upon his return to London from Japan in
1899, Barton-Wright set up a martial arts school to teach Bartitsu
to Englishmen. Presumably that's how a Londoner such as Sherlock Holmes would have learned the technique.

The concept struck a chord with a great number of English
gentlemen, and soon there were many places of Bartitsu instruction. During Britain's Victorian Age canes and walking sticks were
extremely popular and, logically enough, Bartitsu techniques
often incorporated canes. It was reputed that Bartitsu masters

using a light cane or even an umbrella could beat away ruffians brandishing cudgels, shillelaghs, bonkers, batons, bludgeons, or even truncheons.

Eventually the need for such club-based defenses waned, enthusiasm for Bartitsu eroded, and the techniques were nearly forgotten. Fortunately for those interested in living dangerously, there has been a small but enthusiastic recent revival in the practice. Several books and videos on the subject are available from Internet-based sources.

Edgeworkers, like Holmes, may find themselves in precarious situations from time to time and may need to rely on themselves for protection. In those situations, knowing a bit about Bartitsu stick fighting would be a distinct advantage. Logically, the first step is to make a suitable stick or, better put, a "Bartitsu staff."

Making a Bartitsu Staff

There are few hard and fast rules regarding making a Bartitsu staff, as the endeavor is more one of selection than manufacture. Still, there are several considerations.

Style is primarily a matter of preference, as Bartitsu sticks or staffs encompass a wide range of lengths and styles. A cane extends from the wrist to the floor and frequently includes a perpendicular handle or crosspiece at the top. It is suitable for general walking about and, if necessary, support. Canes, while a frequent fashion accessory for the able-bodied in Barton-Wright's day, are now more often used by those who need assistance in walking and standing, the aged, and the infirm.

Able-bodied people use a hiking staff to negotiate inclines and speed passage in hilly terrain. A hiking staff extends to the shoulder and substitutes a keeper or thong for the perpendicular handle of the cane. This is easy to make and you might find it more useful than a cane. The hiking staff works well for Bartitsu.

Materials and Tools

- Handsaw
- Straight sapling with an average diameter of roughly 1 to 1½ inches, and long enough to reach, after cutting, from the top of your shoulder to the ground. (The best woods for hiking staffs include ash, hickory, oak, and ironwood. Other types such as maple and birch may be used as well. Be sure to secure permission from the landowner before you cut down any saplings.)
- Knife suitable for removing bark
- Medium grit sandpaper
- Stain and varnish, your choice
- Drill
- ³⁄₁₆-inch drill bit
- (2) 12-inch long leather thongs

Step 1. Using a handsaw, cut the staff to length and then let it dry in a covered place, standing up, undisturbed for at least four weeks.

Step 2. With a knife, carefully remove the bark, knots, and uneven spots from the staff surface. Smooth the staff with sandpaper. Stain and varnish the staff as desired. Let it stand until the varnish is completely dry. Refer to the package directions for drying time.

Step 3. Starting 2 inches down from the top, drill three ³⁄₁₆-inch-diameter holes through the staff approximately 9 inches apart. Insert a 12-inch-long leather thong through the top hole and tie the ends with a square knot. This is the keeper and goes around the user's wrist when hiking. Insert a second 12-inch leather thong through the bottom holes and tie each end off with a double overhand knot. This is the sling and allows the hiker to carry the staff on the shoulder when both hands are needed for negotiating uneven terrain.

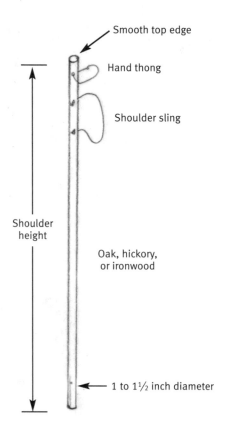

Smooth top edge

Hand thong

Shoulder sling

Shoulder
height

Oak, hickory,
or ironwood

1 to 1½ inch diameter

Bartitsu staff

Bartitsu Style Self-Defense with a Hiking Staff

You can use a staff, umbrella, or cane to great advantage if you ever
find yourself in a dicey situation. In January and February of 1902,
Barton-Wright published a series of articles in the then-popular
Pearson's Magazine entitled "Self-Defense with a Walking Stick."
Barton-Wright's lengthy article is a useful, if old-fashioned, treat-
ment of techniques that can be used to "defend one's self with a
walking stick or umbrella when attacked under unequal conditions."
(It would work just as well with a longer hiking staff.)

While it takes long practice to truly master any martial art, the Barton-Wright article provides some easily digestible information on Bartitsu-style self-defense that may be useful to edgeworkers on (hopefully) infrequent occasions.

The original articles explained 22 different scenarios including "The Safest Way to Meet an Attack with a Spiked Staff or Long Stick when you are Armed only with an Ordinary Walking Stick," and "How to Defend Yourself if you are Carrying Only a Small Switch in your Hand and are Threatened by a Man with Very Strong Stick." All of the techniques make for interesting, if quaint, reading. I've reproduced one of the articles along with the original *Pearson's Magazine* artwork below, as I believe this represents the most likely situation that a dangerous but artfully living person may face. However, if you're one of those on the far right side of the risk curve—that is, if you frequently find yourself in situations requiring hand-to-hand combat—then it may be worth your time to study the entire article and review all 22 techniques.

One of the Best Ways of Knocking Down a Man in a General Scrimmage, when there is not Room to Swing a Stick Freely.

When a man finds it necessary to defend himself in a street fight, or the like, he may not have room to swing a stick freely. One of the best methods of using a stick as a weapon under these circumstances is to pass it between the legs of the assailant, and, by pressing it sharply against the inside of one of his thighs, to cause him to lose his balance.

In order to carry out the trick effectively on a single assailant, when there is no crowd, you should stand in the front guard position, and make a cut at the side of your opponent's face. While he raises his hand to guard his face, you seize his uplifted hand with your left hand, crouch down and pass your stick through his legs, exerting sufficient leverage to throw him on his back.

Another method is to take up the back position guard, standing with your left foot forward and your right arm above your head, which you

Bartitsu

must purposely expose in order to induce your opponent to strike at it. At the moment when he attempts to hit you on the head, you must slip under his guard, and seize his right wrist. Now pass your stick between his legs, and throw him upon his back.

To employ the same trick in a crowd it is only necessary to stoop, cover your face well with your arm and hand, and to keep diving with your stick between people's legs, upsetting them right and left.

12

Thrill Eating

"Der Mensch ist was er isst." *(Man is what he eats.)*
—LUDWIG ANDREAS FEUERBACH,
GERMAN PHILOSOPHER, 1863

Big-T personality types often have not only a symbolic but also a literal appetite for danger. If you are what you eat, then eating dangerously is a trait worthy of developing and refining.

Big-T foodies travel the world in search of food-related experiences they hope will lead them down pathways to enlightenment and spiritual growth. These risk-loving personality types can often fight through the natural disgust they feel at consuming unfamiliar and vile tasting foods. One of Freidrich Nietzsche's most memorable quotations is "what doesn't kill you makes you stronger." This may be at the core of what motivates novelty seekers to attempt things that take them so far from their comfort zones.

A Plateful of Danger

Consider the hardy (or is it foolhardy) gastronomes who travel to the Mediterranean island of Sardinia in search of the forbidden cheese, the Casu Marzu. Casu Marzu is so dangerous and unhealthy it is illegal to make and consume even in Sardinia, evidently the one spot on planet Earth where people actually enjoy the taste. The cheese is a specially prepared variety of a sheep's milk cheese called Pecorino Sardo. However, obvious differences separate the delightful, biting taste of Pecorino Sardo and the less delightful half-inch-long biting maggots infesting Casu Marzu.

To make the cheese, the clandestine cheesemaker purposefully mixes live fly larvae of the genus *Piophila casei* into the nascent cheese during its fermentation. The ripe cheese contains live, wriggling maggots that can pass undigested into the intestine, where they place the eater at risk of severe intestinal distress. The taste and texture of the cheese, noted Yaroslav Trofimov in the *Wall Street Journal,* is "a viscous, pungent goo that burns the tongue."

But it's not just what you eat that makes for dangerous living, it's how you eat it. The International Federation of Competitive Eating organizes and sanctions a number of well-known eating contests. The biggest events include New York City's Nathan's Hot Dog Eating Contest, Philadelphia's Wing Bowl, and the Chattanooga Krystal Square Off World Hamburger Eating Championship. Competitive eating is a strange world, often dominated by small-statured Asian men consuming a sizeable proportion of their body weight in a single meal. Superstar super-eater Takeru Kobayashi, the Michael Jordan of competitive eating, has consumed as many as 63 hot dogs (with buns) in 12 minutes. He weighs just 160 pounds fully dressed.

There is an element of outlier-style danger in competitive eating. Due to a rash of bad outcomes, the *Guinness Book of World Records* will no longer accept or recognize eating-related records.

In all of these activities danger definitely exists, but does that special sense of artfulness reside there as well? Most people would agree that existing on a high-calorie, high-fat diet, as Morgan

Spurlock did for the movie *Supersize Me,* does not constitute artfully dangerous eating. And it is easy to agree that the knowing
consumption of tainted or spoiled food has no place in anybody's
diet. Gastronomic adventurers such as Anthony Bourdain and
Andrew Zimmern have written and broadcast extensively about
sampling very, very strange foods, often involving arachnids,
crustaceans, insects, and other arthropods.

Zimmern, who hosts the national television show *Bizarre Foods
with Andrew Zimmern,* in particular has made a career of eating dangerously. Among his favorite experiences is Hakarl (rotted Ice-
landic shark) and Ecuadorian guinea pig. Andrew was eager to
share a number of excellent tips with kindred spirits—edgeworkers
who choose to live and eat within the Golden Third.

Andrew Zimmern's Top Tips for Eating Dangerously

- First, eschew your psychological barriers of custom and
  preconceived notions and embrace the thrill. You need to
  get past the idea that something foreign is bad for you.
- Check out the street food. When I'm in a new place, the
  first place I go is to the street vendors. And not the main
  street, but the street behind the street behind the street.
  That's where the food is the best and most authentic.
- To figure out where to eat on the street, use what I call the
  New York City hot dog school of good eating. Imagine
  you're outside the New York Public Library and there are
  five hot dog vendors. Four of them look grouchy; they have
  their hands in their pockets and there's no line. But one
  guy has a long line, he's smiling, and his customers are
  smiling. That's the guy to buy from.

- Figure out where the cabbies eat. The cabbies and people like them eat on the cheap. They aren't "dining," just eating, but they do it a lot and know what's good. If you go to those same places, you may get into some really exciting foods.
- Seek out the street markets. It's the first place I check out in any city. I go to the farmer's markets where it's the daily part of life in which everybody participates. It's where you really see what's available, local, and fresh.
- I travel prepared. In certain situations after, say, a long day on the plane, I may coat my stomach with Pepto-Bismol to provide it with some protection. And I carry an antibiotic such as Cipro with me, just in case.

Some food and drink is endowed with a rich, colorful history spanning decades and continents that adds immeasurably to the allure of trying it yourself. There are foods that do have a reputation for danger, deserved or not. Make a minor error in the preparation or consumption techniques for these foods and drinks and you can risk catastrophe—ranging from discomfort to madness to death.

Because of their (purported) danger and unique reputations, consuming these foods earns the eater genuine street cred and provides a dependably engrossing cocktail party conversation topic. Fugu is one of these fabled foods.

Fugu

Two hundred years ago, Japanese haiku poet Yosa Buson wrote:

I cannot see her tonight
I have to give her up
So I will eat Fugu.

Fugu is the Japanese name for the blowfish or tiger pufferfish. Eating it is an expensive, sublime treat. When cut expertly and served fresh, it is, as the Japanese say, *shiko-shiko* and *tsuru-tsuru* in the mouth: the perfect combination of chewiness and smoothness. Fugu connoisseurs describe the taste as "subtle as the fragrance of spring rain dripping upon a stone."

Japanese gourmets notwithstanding, not many animals eat the tiger pufferfish, because over the eons of time, this fish has evolved a very effective defense. The natural defenses of fugu are twofold. When threatened, the puffer inflates its stomach, becoming a large, threatening pincushion of a fish. If that's not enough to dissuade the predator, then the puffer submits to its fate, knowing that tetrotoxin, a poison 1,200 times more potent than cyanide, lurks inside its liver and other organs and will soon cause grievous and everlasting harm to the foolish predator that ate it.

Such a defense is perhaps too late to protect this particular pufferfish, but the message its sacrifice sends ultimately provides the puffer's descendants a better chance for success in the gene pool regardless of the size and toothiness of nearby sharks and barracuda.

The great exception to the do-not-mess-with-me rule for pufferfish is the human sushi enthusiast who delights in its unique and unusual taste. Eating the fish is a gamble, for one needs to invest his life into the hands of a person he doesn't know; wagering the fugu chef has the Heifetz-like finger dexterity, knowledge, and complete presence of mind to remove all traces of the deadly poison. Yet a truly superior sushi chef can leave a microscopic trace of the toxin within the preparation, giving the Big-T gourmet a whiff of the sublime flavor of death on the tongue.

Lest one believe the danger is imaginary, mere hype cooked up by braggart foodies to impress their friends, consider this: hundreds if not thousands of Japanese diners have unexpectedly cashed in their chips, their last meal becoming just a bit too transcendent.

Is it possible that sashimi made from the flesh of the tiger pufferfish is available for Big-T foodies in the United States—a place sometimes derided for its compensation culture in which the injured too often seek compensation even when the risk was known in advance and willingly accepted? Indeed it is, and for this, Golden Third adventurers can thank New York City chef Nobuyoshi Kuraoka.

In the 1980s Kuraoka began his campaign to bring fugu to his sushi restaurant in Manhattan. It took more than four years and countless meetings with government officials, but perseverance paid off in this case.

After much negotiation and planning, the FDA gave Kuraoka permission to import blowfish for consumption, pursuant to a few special conditions. All fugu sold in America arrives frozen in sealed plastic containers directly from a single processor in Yokohama, Japan. There in Japan, experts clean and package the fish, freeze it, and then send it by air to JFK airport, where it is inspected and wholesaled to one of 20 or so restaurants in the United States that serve it.

Many sushi restaurants in America purport to serve fugu, but some play fast and loose with the definition. Some restaurants, particularly those not near New York, serve fugu cut not from the prized tiger pufferfish but from the more common and nearly tetrotoxin-free Atlantic pufferfish. This fish provides a much tamer eating experience. It's fairly common in East Coast restaurants, where it's called *sea squab*.

The only restaurants allowed by federal law to serve the real McCoy are those that obtain their fish via Kuraoka's fugu facility in Yokohoma. If sampling the genuine article is important to you, check its provenance carefully before ordering.

Given how extravagantly deadly this poison is, and further given that everybody who prepares it, sells it, and eats it knows that, do people really die from it? Absolutely, but the risk seems to be less now than in the past.

The heyday of pufferfish poisoning was the 1950s. In 1958, says the *New York Times*, more than 150 Japanese gastronomes died from eating fugu. Since then, fugu has dispatched considerably fewer each year, but when playing gastronomic Russian roulette, the ball will still drop into the unlucky hole on occasion.

Mitsugoro Bando VIII, the top Kabuki actor in Japan, lost his bet on January 16, 1975, after a meal in one of the country's top fugu restaurants. If Bando had limited himself to fugu sushi, he might not have become the world's best-known fugu victim. But Bando had a particular preference for soup made from the liver of the fish. And it's in the liver that the tetrotoxin accumulates.

Echoing Beatrice in Nathaniel Hawthorne's story "Rappacini's Daughter," the most daring fugu buffs attempt to build up resistance to the toxin by ingesting small quantities, increasing the measure slowly over time. Bando often ate small bowls of soup made from the liver of the fish. One night, while dining with friends, Bando enjoyed a bowl of fugu liver soup. His friends, afraid of being poisoned, declined theirs. Bando impulsively ate their servings as well.

That evening, Bando told his wife he was feeling wonderful, almost as if he were floating on air. The irreversible poisons in fugu took their time when Mitsugoro Bando VIII laid down his chopsticks for the final time. During the night he did float away, carried to kabuki heaven on the fins of a pufferfish. Death came slowly. First he probably experienced an intense tingling of fingers, toes, and lips. Then his body parts became more and more sluggish, his respiration slowed then ceased, and finally the electrical pathways of his brain shut down.

Ackee

In contrast with fugu, a death caused by eating improperly harvested ackee would not be nearly so peaceful.

Ackees are fruit-bearing trees much beloved in Jamaica, Haiti, and parts of West Africa. The fruits are delicious, spongy, and nutritious. When combined with salted fish, ackee may be the most popular dish in Jamaica. But the fleshy pods of unripe ackee contain hypoglycine, a lethal toxin. Ackee pods hang in clusters amid glossy green leaves, turning bright red as they ripen, finally popping open as if to announce to the world that now they are fine to eat. But they are decidedly not safe to eat until they are ripe.

Unlike fugu, preparing ackee safely is not difficult—nearly anyone can tell a ripe from an unripe ackee if they know what to look for. But the problem is that sometimes they don't. Or they truly love the taste of unripe ackee. Like Bando and his fugu liver soup, some ackee connoisseurs enjoy the danger and believe it causes the food to taste better. It may be possible to build up a tolerance, but that's a thin line and a dangerous calculus.

The ackee is a member of the soapberry plant family, a first cousin to the lychee. Its scientific name is *Blighia sapida,* in honor of William Bligh, the captain of the HMS *Bounty,* who is thought to have introduced it to Jamaica in the late 18th century.

The name for the nasty condition caused by ingesting unripe ackee is Jamaican Vomiting Sickness. By all accounts, Jamaican Vomiting Sickness is a bad way to go. The *British Medical Journal* says JVS begins with abdominal discomfort two to six hours after eating unripe ackee fruit. Next comes severe vomiting, a period of weakness, and then an even worse round of retching. Depending on the amount of toxin ingested and the overall strength of the victim, the terminal phase of Jamaican Vomiting Sickness involves drowsiness, twitching, convulsions, and finally death.

Sampling the delights of the ackee plant is much more accessible and less expensive than dining on Casu Marzu or fugu. The easiest way is to seek out a good Jamaican restaurant in any major city and order ackee and saltfish, a commonly available and very tasty menu item. If you stay away from suspicious restaurants in a Kingston shantytown, there's much less danger involved in eating ackee. But then again, there's always a chance that something could go wrong.

Danger Dogs

It's easy to visualize and accept the peril lurking within the liver and ovaries of a Japanese pufferfish or contained inside the immature arils of Jamaican ackee. But edgeworking gourmands know that gastronomic brinkmanship is also found in more domestic and accessible fare.

When Los Angelenos occasionally leave their cars and walk within the 90 square blocks of traffic-snarled downtown LA called the Fashion District, they often look for the furtive, unlicensed, uninspected street vendors selling a delicious but illegal lunch. Locally it's called the *danger dog* or sometimes the *heart-attack dog*. A typical rendition consists of a hot dog wrapped with grilled bacon covered with a generous glop of a spicy slurry consisting of tomatoes and onions mixed with ketchup, mustard, mayo, and spices. The vendor tops the whole thing off with a large poblano chile, wraps it in a napkin, and says, "Gracias."

Originally conceived on the mean streets of Tijuana, Mexico, the danger dog occupies an iconic place in the pantheon of street vendor cuisine. These sidewalk prepared bacon-wrapped hot dogs are delicious, satisfying, and inexpensive. They are also illegal. The people who make and sell them, typically wildcatting Hispanic street vendors known as *ambulantes*, have been arrested and jailed for selling them. LAPD officers are vigilant in their quest to locate and arrest unlicensed street vendors. Sometimes the vendors spot the cops in time, running away and abandoning their half-full grills of bacon-wrapped contraband. Sometimes they don't and they wind up in jail, in some cases for as long as four months.

While the nutritional value of the dish is small in comparison with the amount of saturated fat and nitrites it contains, that's not the reason it is banned. It's because the health department will not allow hot dogs to be grilled on the street. The legal method of preparation is steaming or boiling. Since you can't steam or boil bacon, the danger dog cannot be sold legally on the streets of Los Angeles.

Perhaps this is merely a case of persnickety overregulation, the result of little-t minded municipal bureaucrats drinking too deeply from the well of government power. But maybe *ambulante*-prepared, bacon-wrapped hot dogs really do represent a legitimate threat to the health and safety of the public.

It's hard to say, but the important question is this: are they worth trying? I give no guarantee or advice regarding food safety, but for those living their lives in the Golden Third, a one-off taste test seems a reasonable risk. Cooked long enough and at a high enough temperature, most germs should be eliminated even under the less than sanitary conditions surrounding their preparation. And one can always adopt a New York City approach: get in the longest line and avoid the unhappy, lonely vendor.

If you can't locate a wildcat hot-dog vendor but want to sample the dangerous deliciousness anyway, here is a recipe from downtown Los Angeles for a heart-attack dog.

Make a Danger Dog

Materials and Tools

- Charcoal grill
- Charcoal chimney or lighter fluid
- 2 medium onions, sliced
- 6 medium poblano peppers, seeded
- 6 hot dogs
- 6 slices of bacon
- Wooden toothpicks
- 2 medium tomatoes, diced
- Ketchup, mustard, mayonnaise, salt, and pepper

Step 1. Place onions and pepper directly over the grill.

Step 2. Grill hot dogs until a light char crust forms on the surface.

Step 3. Wrap a piece of bacon around each hot dog and secure with a toothpick. Grill the wrapped hot dog until the bacon is crispy, turning the dog at intervals.

Step 4. Place the hot dog in a bun and cover with tomatoes, grilled onions, and condiments, and top with the grilled pepper. Remove the toothpick and serve.

Bhut Jolokia

Can eating chile peppers be dangerous? Well, if by dangerous you mean life threatening, then probably not, at least in most cases. I've surveyed the literature on the subject and found only a few cases of chile pepper lethality. Most claims of DBCO (death by chile overdose) are dubious at best, although there does seem to be at least one fairly well-documented case in India. But there are numerous cases of fainting and spontaneous vomiting caused by an ingestion of a cracker with a little too much super hot sauce spread on it.

In past times, chiles were an ingredient in the poison on the dart tips of Borneo jungle hunters. In China, they were rubbed into sensitive body parts as a form of punishment and torture. The *Codex Mendoza,* a 1541 illustrated manuscript describing contemporary Aztec life, contains a rather disturbing illustration of an Aztec boy being held in the smoke of burning chiles as punishment for some childish infraction.

The physiological effect of peppers upon the human senses is complex. Although we use the word *burn* to describe what we feel, the burn of chile pepper differs from a temperature burn or a chemical burn. When a hot chile is eaten, the brain immediately reacts to the capsaicin in the chile, stimulating nerve receptors and releasing chemicals into the bloodstream. The pain is the result of direct stimulation of a pain-sensing neuromechanism manifested by an intense burning sensation in the mouth, tears in the eyes, and copious sweating.

Here is the burning question: what is the most dangerous chile? Or put another way, what is the world's hottest chile? The Scoville scale measures the hotness or, more correctly, piquancy of a chili pepper. The units of the scale are Scoville heat units (SHU) and they indicate the amount of capsaicin present.

The scale is named after its developer, American chemist Wilbur Scoville. His method, which he devised in 1912, is known as the Scoville Organoleptic Test. The scale runs from 0 (no capsaicin) to 15 million (100 percent capsaicin).

SCOVILLE HEAT UNITS	PEPPER TYPE
15,000,000	Pure capsaicin [5]
2,000,000–5,300,000	Standard U.S. Grade pepper spray
855,000–1,041,427	Bhut Jolokia
350,000–577,000	Red Savina Habanero
100,000–350,000	Habanero Chili, Scotch Bonnet
100,000–200,000	Jamaican Hot Pepper, African Birdseye
50,000–100,000	Thai Pepper
30,000–50,000	Cayenne Pepper
10,000–23,000	Serrano Pepper
5,000–10,000	Hot Wax Pepper
2,500–8,000	Jalapeño Pepper
1,000–1,500	Poblano Pepper
500–2500	Anaheim Pepper
100–500	Pimento, Pepperoncini
0	Bell pepper

For years, the Red Savina Habanero was considered the world's hottest pepper. Now, a new one has been discovered. It's so hot it makes your typical habanero look like a cinnamon stick. Called the Bhut Jolokia (also Naga Jolokia, Bih Jolokia, Ghost Chile, or the Poison Pepper), it grows in northeastern India and Bangladesh. In 2007, Guinness World Records confirmed Bhut Jolokia as the hottest chile, surpassing the Red Savina by a whopping 300 percent. This pepper is so powerful it can stop a charging elephant.

Wildlife experts in northeastern India are experimenting with a new weapon to prevent marauding elephants from destroying homes and crops—superhot chilies. Conservationists working on the project in Assam state said they have put up jute fences made of strong vegetable fiber and smeared them with automobile grease and bhut jolokia chilies.

—THE ASSOCIATED PRESS, NOVEMBER 20, 2007

Sampling the Bhut Jolokia, or preferably, a sauce made from it, is definitely on the must-do list for those who eat in the Golden Third. Finding it may not be easy, however, growing as it does in the far-off provinces of India. But gardeners have access to seeds from Internet-based seed growers, giving them the opportunity to try the pepper if they have a green thumb to go with their iron stomach. Typing "bhut jolokia seeds" or "naga jolokia seeds" into an Internet search engine should provide contact information for several vendors.

Albuquerque's Dave DeWitt, aka the Pope of Peppers, literally wrote the book on chile peppers. DeWitt has written more than a dozen books on chiles, and his *Chile Pepper Encyclopedia* is the authoritative tome on all things chile pepper related.

Oleoresin capsicum is the concentrated heat ingredient in super-hot sauces with names using words like "insanity," "death," and "suicide." Most people react very negatively to the super-hot sauces, experiencing severe burning and sometimes blistering of the mouth and tongue. Other reactions may include shortness of breath, fainting, nausea, and spontaneous vomiting. People should be very careful of commercial hot sauces that list oleoresin capsicum as an ingredient, and taste them in small quantities.

—*DAVE DEWITT*, CHILE PEPPER ENCYCLOPEDIA

Dave DeWitt's Top Tips for Enjoying Hot Chile Peppers

- Build up your tolerance by starting with milder chiles. As you grow more comfortable with piquancy, you can move up the Scoville scale to hotter ones.
- Drinking water is ineffective in relieving the burning sensation capsaicin causes. Capsaicin is not soluble in water.

Water merely spreads the oils throughout your mouth. Milk, yogurt, and sour cream are more effective at relieving piquancy discomfort. Research indicates that the casein protein found in dairy products dilutes or encapsulates the offending chemicals.

- You can diminish the burning sensation by thoroughly brushing and rinsing your mouth, especially the tongue, with a soft toothbrush and regular toothpaste.
- The spiciest parts of peppers are the placental tissues, that is, the stringy interior that surrounds the pepper's seeds. Avoiding them will typically decrease the piquancy.
- Wash your hands after handling peppers and before touching any sensitive body parts (for example, when using the bathroom or inserting contact lenses).

13

Flamethrowers

Nothing attracts the attention of those within the Golden Third quite like a flamethrower. I've placed this material at the back of the book for a reason: it's something to build up to.

The first time I saw a flamethrower in action was when my friend Christian Ristow brought his flamethrower-equipped robot to a midnight machine art festival in a Phoenix junkyard. My son, Andy, was with me, and neither of us will ever forget the roar and the light and the heat of the 20-foot-long tongue of flame that Ristow's flamethrower could project.

Now I have my own flamethrower, and it generates a frisson of excitement in all who see it work. Better yet, it fills me with pride at my ability to safely craft and use something like this. Much of what I include in this chapter, I learned from alumni of the robotic art group called Survival Research Labs. This San Francisco–based ensemble pretty much wrote the book on fire-based performance art. SRL has had its ups and downs, having been banned for life from several countries and kicked out of several cities—mostly for violating local fire codes with perform-ances that invariably include giant robots sporting giant flamethrowers.

Aside from artistic pretensions, flamethrowers have their practical side. Flame-based weaponry is among the oldest military hardware known. The use of incendiaries in battle dates back to Classical Greece, if not further. Historian Thucydides wrote in the *History of the Peloponnesian War* about a giant flamethrower built by the Boethians. It consisted of a "great beam sawed in two and scooped out and fitted together again like a pipe." From it, the Greeks shot a mixture of "lighted coals, sulfur, and pitch, which set fire to the fortress wall making it untenable for its defenders, who fled."

But the great breakthrough in flamethrower technology occurred when the Byzantine Empire perfected its ship-borne flamethrower. Attached to the bow of a trireme warship, the Byzantine flamethrower was a metal tube and bellows. The Byzantine Greeks used the projector to shoot a nearly unquenchable burning goo they called Greek Fire. The Greeks kept the exact formula a secret, but many historians speculate that the stuff was made from a mixture of petroleum, pitch, and sulfur.

Flamethrowers have been used, off and on, in military campaigns since that time, notably by the German army in World War I in the Belgian trenches and by American soldiers in the Pacific during World War II. Those military flamethrowers were typically carried on a soldier's back. The backpack consisted of two cylinders, one holding a compressed gas and the other a flammable liquid. When the flamethrower operator pressed the trigger, the gas forced the liquid out of a nozzle, igniting the liquid as it passed through, and spraying a 100-foot radius with death and destruction.

Flamethrowers have a less lethal practicality as well. They are frequently used in agriculture, typically for tasks such as clearing weeds and brush between rows of crops and for sanitizing chicken coops. There also is a variant called a propane cannon that makes a very loud noise, and it typically is used to scare birds away from airport runways and berry fields.

Military flamethrowers shoot a flaming liquid such as napalm, while commercial and agricultural ones are propane based. While neither type can be considered intrinsically safe, propane devices are less dangerous. Of course, danger is relative. As far as I'm concerned, liquid-based flamethrowers should be kept in the same part of the garage as your plutonium, spitting cobra, and vial of Ebola virus.

The non–liquid-fueled, propane flamethrower described here, technically called an accumulator-based flame cannon, is a project falling within the Golden Third—presuming you approach it with adequate care and preparation. That preparation includes reading and understanding what follows.

The Flamethrower

Before we get started with flamethrower fabrication, decide if you really want to do this. I think flamethrowers are novel and inter-esting, but you may feel differently. The risks (serious burns, set-ting the garage on fire, or a visit from an angry policeman) may not be worth the return. If that's the way you feel, no worries, just move on to a different project.

But if you know this and still want to go ahead, then read on.

Start by reviewing chapter 5. I've made several flamethrowers with no major problems. But remember, things can go wrong even through no fault of your own. If you do attempt this project, you and you alone are responsible for what happens.

Flamethrowers are showy, dangerous, exciting, impractical, and visceral. In short, they epitomize the art of living dangerously, but only if made well and used responsibly. Go ahead if you dare, but only after heeding the warnings below.

Safety Guidelines

- Do not operate the flamethrower near combustible materials. Keep people and animals away.
- Inspect equipment for damage and wear prior to each use.
- Use only nonmodified, government-approved propane cylinders.
- Keep the propane cylinder level and upright. Don't invert cylinders or lay them on their sides. Cylinder valves must be protected. Never lift a propane cylinder by the valve.
- Don't use a flame to heat up a gas container in order to increase pressure.
- Shut everything down if you smell gas. Immediately shut off all valves. Never use a flame to test for leaks. Instead, use soapy water and look for bubbles.
- Propane is heavier than air, so it will accumulate in the nozzle and other bowl-shaped or low areas. Be certain your area is well ventilated.
- Keep all sources of ignition away from cylinders, regulators, and hose.
- Wear protective gear including safety glasses, leather apron, and heat-resistant gloves.
- Have a fire extinguisher close at hand.
- Comply with all safety guidelines and local ordinances regarding the use of an open flame.
- Use extreme caution at all times. You are using an intense open flame, and disregarding safety practices can have severe consequences. For Pete's sake, *this is a flamethrower.*

Materials

Accumulator Assembly

- (1) 20-pound standard government-approved propane cylinder, filled. Such cylinders are widely available and are used to fuel propane grills.
- (1) Variable-setting, high-pressure propane regulator with a 10-foot hose.
- The propane regulator used by most backyard barbeque grills is a fixed-pressure, low-pressure device and will not allow sufficient pressure to accumulate for our purposes. Instead, obtain the type of regulator associated with the large burners used on turkey fryers and much larger commercial grills. To find one, use the following term in any Internet search engine: "high pressure propane regulator." The best regulators allow you to set the pressure by turning a valve. Some regulators allow pressures as high as 50 PSI, but don't exceed 30 PSI.
- (1) $\frac{1}{2}$-inch-diameter steam whistle valve (often available online through eBay and other sources. Internet search term: "steam whistle valve.")
- (1) 2-inch-diameter black iron (BI) pipe, NPT threaded both ends, 24 inches long
- Miscellaneous reducing fittings to reduce from the 2-inch threaded pipe to a $\frac{1}{2}$-inch-diameter NPT female thread on both ends. You need a combination of reducing couplings and pipe nipples. A typical collection of fittings is shown in the assembly diagram, although other combinations of fittings could work.
- (1) $\frac{3}{8}$-inch diameter flare fitting to $\frac{1}{2}$-inch diameter NPT fitting, male both ends
- Gas-rated pipe thread sealing compound
- 15 feet of rope

Stand Assembly

- (2) 2-inch-diameter black iron (BI) pipe nipples, 4 inches long
- (1) 2-inch-diameter BI pipe tee fitting
- (1) 2-inch-diameter BI floor flange
- (4) $\frac{1}{4}$-20 bolts, 2 inches long
- 3 foot x 3 foot plywood base. (You may stake or weight the base if necessary to make certain it will not tip during operation.)

Materials (Continued)

Nozzle Assembly
- (1) ½-inch-diameter pipe nipple (nozzle holder), 5 inches long
- (1) ½-inch-diameter to

¾-inch-diameter coupling (the nozzle)
- Medium-mesh steel screen

Igniter Assembly
- Disposable steel propane bottle (BernzOmatic is a popular brand.)
- Propane torch to fit bottle
- (1) 9 by 12-inch (approximate) heat-resistant cloth (The Model

HC9X12 BernzOmatic Heat Cloth is one brand name.)
- (2) 6-inch-diameter steel hose clamps
- Flint and steel spark igniter made for propane torches

Tools

- Pipe wrench
- Drill
- #7 drill bit
- ¼-20 tap
- Miscellaneous workshop tools including pliers, hammer, screwdrivers, and wire clippers

- Safety glasses
- Heat resistant gloves (welder's gloves)
- Leather welding apron
- Heat resistant welder's cap

Step 1. Build the Flamethrower Assembly

Lay out all the pipe and pipe fittings on a table.

Begin by screwing together the lower section of the 2-inch pipe fittings and pipe that forms the accumulator tank.

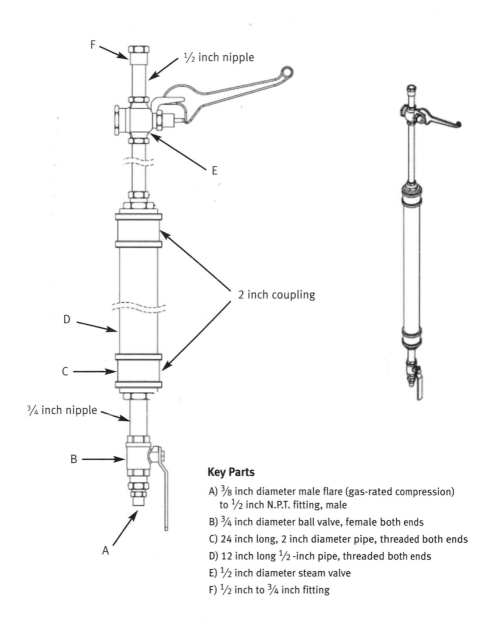

Key Parts

A) $3/8$ inch diameter male flare (gas-rated compression) to $1/2$ inch N.P.T. fitting, male

B) $3/4$ inch diameter ball valve, female both ends

C) 24 inch long, 2 inch diameter pipe, threaded both ends

D) 12 inch long $1/2$-inch pipe, threaded both ends

E) $1/2$ inch diameter steam valve

F) $1/2$ inch to $3/4$ inch fitting

Flamethrower assembly diagram

Starting at the top of the 2-inch pipe, attach pipe reducing fittings that change the diameter of the pipe from the 2-inch pipe forming the accumulator tank to the ½-inch-diameter pipe that feeds the steam valve. The assembly diagram shows the relative position of the pipe and pipe fittings. As noted in the materials section, you will need to select pipe fittings that allow you to transition from the 2-inch-diameter accumulator pipe to the steam valve. Use gas-rated thread compound on all joints.

Attach a 15-foot-long rope to the steam valve handle.

Attach the ½-inch-diameter pipe and the pipe fitting that forms the flamethrower's nozzle to the accumulator tank. Apply gas-rated pipe compound as you screw the pieces together. Read and follow label directions and ensure that the pipe threads are well sealed.

Step 2. Build the Flamethrower Stand

Drill and tap holes to accommodate the four 2-inch-long ¼-inch diameter bolts in one of the 2-inch-diameter nipples in the location shown in the flamethrower diagram. Use a #7 drill bit and a ¼-20 tap to make the threaded bolt holes in the pipe nipples. Tapping a hole in black iron is not difficult, but if you've never done it before, look up the proper procedure before attempting.

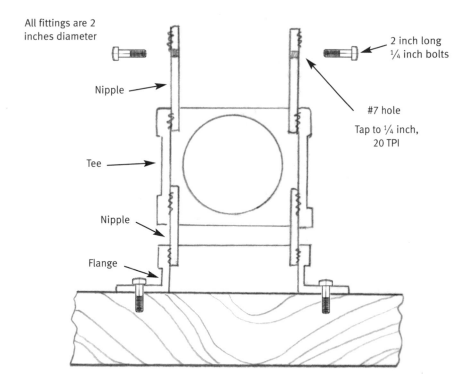

All fittings are 2 inches diameter

2 inch long ¼ inch bolts

Nipple

#7 hole
Tap to ¼ inch, 20 TPI

Tee

Nipple

Flange

Flamethrower stand diagram

Assemble the flamethrower stand by screwing together the stand assembly pieces listed in the materials section. After the pipe is assembled, securely attach the floor flange to the wood base with the ¼-inch bolts, washers, and nuts. Refer to the flamethrower stand diagram as needed.

Step 3. Install the Igniter Assembly

Screw the propane torch onto the small propane bottle. Wrap the small propane bottle in the protective fireproof cloth. Using the two hose clamps, wrap the bottle with the heat-resistant cloth and affix the propane bottle/torch so that the torch nozzle extends 1 inch directly above the flamethrower nozzle.

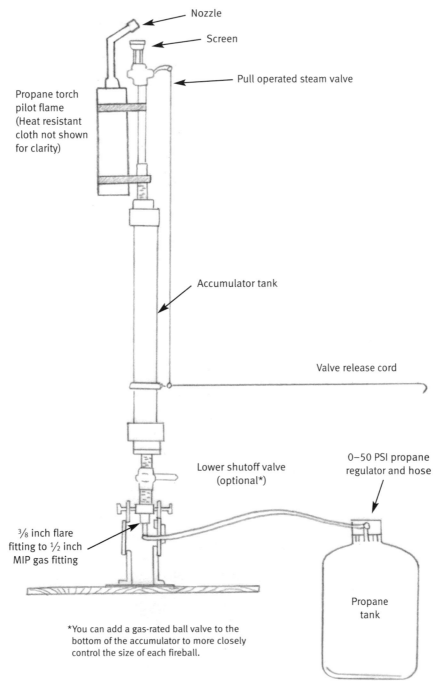

Nozzle

Screen

Pull operated steam valve

Propane torch
pilot flame
(Heat resistant
cloth not shown
for clarity)

Accumulator tank

Valve release cord

0–50 PSI propane
regulator and hose

Lower shutoff valve
(optional*)

³⁄₈ inch flare
fitting to ½ inch
MIP gas fitting

Propane
tank

*You can add a gas-rated ball valve to the
bottom of the accumulator to more closely
control the size of each fireball.

Flamethrower igniter diagram

Step 4. Install the Spark Arrestor

Cut a round piece of wire screen approximately the same size as the nozzle. Bend the screen (used here as a rudimentary spark arrestor) slightly if necessary and make it fit snugly under the lip of the reducing coupling.

Step 5. Connect the Regulator

Connect the high-pressure regulator to the propane tank. Thread the hose through the opening on the stand and connect the hose to the ⅜-inch-diameter male flare fitting on the propane hose to the female fitting on the end of the flamethrower handheld assembly.

Insert the flamethrower assembly into the flamethrower stand. Center the flamethrower in the stand as vertically as possible and secure by turning the positioning bolts tapped into the stand.

Flamethrower gas connections

Step 6. Test the Assembly for Leaks

Turn on the gas valve on the main gas cylinder and test the assembly for leaks by checking all connections with soapy water. Bubbles indicate leaks. Repair any leaks prior to using flamethrower. If you smell gas, stop and fix the leak. After testing, turn the valve off.

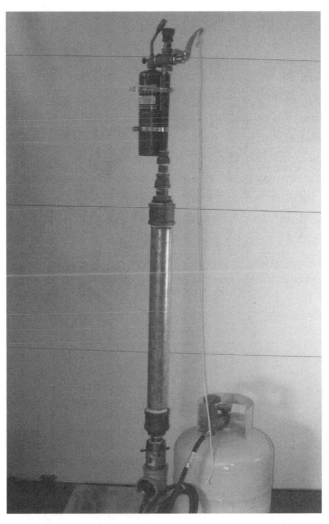

Flamethrower completed (heat-resistant cloth not shown for clarity)

The standard gas bottle used to fuel barbeque grills is filled with liquid propane. When the gas bottle valve is opened, the propane boils and turns into a gas, which is ignited by the pilot flame. The accumulator tank is necessary because without it the vaporizing propane would ice up the flamethrower, and it would not work.

Step 7. How to Operate (For Adults Only)

Don protective safety equipment including (but not limited to) safety glasses and gloves.

Remove anything flammable from the area and add a reasonable margin of safety.

Operate the flamethrower only in areas suitable for a device of this kind. The area should be secured to keep people and animals from accidentally entering it.

Completely clear the area beneath the flamethrower nozzle for a minimum of 10 feet in all directions. Propane is heavier than air, so non-ignited propane can drift downward from the nozzle and burst into flames suddenly when the propane ignites. Therefore, *the area underneath the nozzle is a danger zone!* Do not stand or put any part of your body underneath the nozzle when you operate the device. Stand well off to the side.

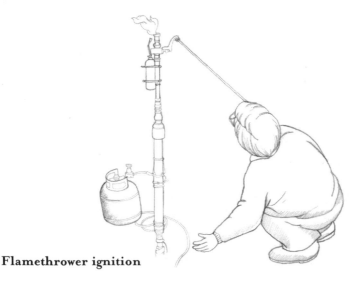

Flamethrower ignition

Open the valve on the small propane bottle and ignite the torch with the flint and steel torch igniter. Step back 10 feet and SLOWLY open the valve on the main gas cylinder. Slowly pull the rope attached to the whistle valve to release a small amount of propane. Once a small flame is present in the nozzle, pull the valve completely open. A large fireball will issue from the nozzle. Note that a small flame, lingering from residual propane in the nozzle, will be present for several seconds. This small flame can be used to ignite the propane for subsequent fireballs. To stop, release the whistle valve.

Flamethrower in action

Step 8. Shutdown Procedure
Follow this procedure to shut down the flamethrower.

    1. Turn off the valve on the propane tank.
    2. Turn off the valve on the propane torch bottle.

3. Lay the flamethrower on its side, so the nozzle is pointing slightly downward, making sure nothing combustible is in front of the nozzle.
4. Open the steam valve.
5. Use the igniter to burn off the residual propane in the flamethrower assembly. When all propane has left the flamethrower assembly, you can disassemble it.

There is a time and place for everything. While the times and places for using your flamethrower may be somewhat limited, a bit of thought and examination should uncover suitable venues.

Actually using a flamethrower of your own creation is a thrilling experience, one that you'll no doubt find exciting and memorable. And it provides uncontestable credentials to all who participate as to your status as a person who understands and practices the art of living dangerously.

14

The Strange Music Starts

"The blast that wrecked my family's Sunday morning 40 years ago remains my most spectacular, and certainly most memorable, involvement in chemical research. At around 8 A.M., as the first weak rays of Glaswegian sunshine stole across my room and illuminated the walk-in cupboard that acted as my laboratory, there was a detonation of Verdun-like proportions. A cloud of ammonia-rich chemicals whipped across my room, spraying out pieces of laboratory glassware. My mother shrieked across the hall; my father raced into my room, cursing with unsuspected fluency; and I leapt from my bed as glittering fragments tinkled around me."

—ROBIN MCKIE, SCIENCE EDITOR,
LONDON OBSERVER

Living dangerously is an art, a learnable and improvable skill that, when done well, enhances life without cutting it short. Living dangerously, artfully, enhances not only personal growth and well-being but also society at large. Many great contributors to the world—scientists, politicians, inventors, and writers—were skilled at dangerous living. Often, they learned the skill early in life.

Intel cofounder Gordon Moore (he of the famous "Moore's Law") set off his first boom in Silicon Valley two decades before pioneering the design of the integrated circuit. One afternoon in 1940, the future father of the semiconductor industry knelt beside a stack of dynamite sticks he had fashioned and lit the fuse. He was 11. Imagine an 11-year-old experimenting with such materials in this century. Such activities would likely land the young person in trouble and probably in juvenile court. But in bygone times, flirting with danger was a right of passage, and one that wasn't necessarily discouraged.

Consider how the parents of Francis Crick, the codiscoverer of DNA, handled their son's appetite for living dangerously. Crick wrote in his autobiography that at the age of 10 or 12 (this would be approximately 1928), "I put explosive mixtures into bottles and blew them up electrically—a spectacular success that quite naturally worried my parents. [But] a compromise was reached. A bottle could be blown up but only while it was immersed in a pail of water." Such instances are mentioned too often in the biographies of the great and eminent to be a mere statistical quirk.

Winston Churchill's cousins loved to spend time with him. As one put it, "We thought he was wonderful because he was always leading us into danger." Indeed, during his long life Churchill experienced numerous brushes with death, many resulting not from the rifles and cannon of enemy soldiers but from his propensity for sensation seeking. As one of his biographers points out, the school-aged Churchill "once devised a bomb with the intention of blowing up a house, which was said to be haunted. Having lit the bomb and lowered it into the cellar, he waited for the explosion. When it was not forthcoming, he looked down to investigate—whereupon the bomb exploded. He was stunned but was fortunate to escape with only a dirty face and singed eyebrows."

Similarly, teenaged Thomas Edison ran a newspaper concession on a Michigan commuter train. In his spare time, he set up a portable chemistry lab in one of the train's baggage cars. One of his experiments went awry and destroyed the baggage car, effectively ending his railroad career. On the other hand, his curiosity and love of living dangerously opened up another career that proved far more lucrative in the long run.

Businessmen David Packard and William Hewlett, sociologist B. F. Skinner, inventor Nikola Tesla, rocket scientist Homer Hickam, steamship inventor Robert Fulton, GE chairman Jack Welch, Russian president Boris Yeltsin, and many others could point (assuming they still had index fingers, which Yeltsin did not) to some episode in their past where something they were working on blew up, perhaps unexpectedly, but often on purpose. Those acts of intellectual and scientific risk taking served them well in later life. How different the world would look if young Gordon Moore had been sent to juvenile hall or if Edison had retired from the railroad after 40 years as a telegraph operator.

Is there a battle taking place between Big-T and little-t worldviews? Big-T's accept reasonable risk as long as it provides new experiences and staves off mental boredom. Little-t's, with their brain chemistry operating in a much different way, eschew any avoidable risk.

The little-t's may be gaining ascendance, for it's becoming harder to legally make and do interesting things, especially edgy, highly kinetic things. For a variety of motivations, ranging from fear of terrorism to fear of litigation to fear of unknown outcomes, it's rapidly becoming harder for those on the right side of the thrill-seeking curve to satisfy their natural and reasonable desires.

Life in the Golden Third is under pressure. Pick up the newspaper or watch the news, and you may quickly get the impression that overcautious teachers, tail-covering insurances companies, and by-the-book civil authorities are doing everything in their power to stifle worldly and scientific curiosity. (For these reasons my lawyer asks me to tell you to consider your location, insurance policy, and general situation carefully before embarking on flamethrower construction.)

We, the intellectually curious, may soon find ourselves trapped in a pen, fenced in by rule-bound sticklerism and overzealous concern for our personal safety, unless we exercise our civil liberties and our curiosity. The intellectually inquisitive and those who choose to live dangerously and artfully must demand their right to control their world. It's time to retake authority from those whose goals are to limit, not expand, intellectual and physical pursuits.

> [As youngsters] we learned a great deal, because chemistry not only lets youthful practitioners make stinks and bangs, it lets them test things and record conclusions with instant—usually gratifying—results. It satisfies youthful inquisitiveness in a spectacular manner.
>
> Or at least it used to, for thanks to a host of health and safety measures that have been introduced over the past decade, the nation's youth is now denied such stimulation. It has become taboo to allow young people access to anything more harmful than a piece of litmus paper: the chemist who sold us sodium would be jailed and the teacher who turned a blind eye to our petty pilfering of his stock would be sacked. And jolly good, too, you might think. Can't have our kids blowing themselves up.
>
> —ROBIN MCKIE, SCIENCE EDITOR,
> LONDON OBSERVER

Putting genetics, biology, and brain chemistry aside, it is clear that one reason people take risks is simple: they are curious. Curiosity is a powerful, visceral need that only gets satisfied at a price. The more curious a person is, the dearer the price he or she is willing to pay. Those in the Golden Third understand life is a risky business and that there is *nobility* in risk. Life without edgework is boring, perhaps even ignoble.

It is not just my opinion that risk taking is central to a well-lived life. Many writers, philosophers, scientists, and others have made this point. And statistical evidence suggests that seeking out novel experiences, trying new things, and avoiding routine makes you, in some ways, a better person. Living dangerously makes you smarter and more productive on the job, more valuable to the people you know, and no doubt a more interesting conversationalist at dinner parties. Most importantly, it makes you feel happy and stimulated and self-aware. This feeling of exhilaration might be what Hunter Thompson termed "the strange music."

> With the throttle screwed on there is only the barest margin, and no room at all for mistakes. It has to be done right . . . and that's when the strange music starts, when you stretch your luck so far that fear becomes exhilaration.
>
> —HUNTER S. THOMPSON,
> HELL'S ANGELS

I hope you hear the music. Good luck. Be adventurous, artful, and alive!

Notes

Introduction

1. For a thorough treatment of the relationship between heredity, brain chemistry, and risk-taking behavior, see *Biological Bases of Sensation Seeking, Impulsivity, and Anxiety* edited by Marvin Zuckerman.

Chapter 1: Big-T People, little-t people

1. Jared Ledgard's book *The Preparatory Manual of Explosives* (Morrisville, NC: Lulu Press, 2007) covers the preparation and use of more than 150 explosive compounds, including mercury fulminate. While it's interesting reading, I think it's highly doubtful that reading a single book could impart enough knowledge to allow a novice to make such powerful explosives safely or successfully. According to Ledgard, "Prepare MF in small quantities only and use great care to avoid shock, percussion, heat and friction. Use extreme caution when handling 90% nitric acid which evolves toxic fumes of nitrogen oxides, use proper ventilation and avoid inhalation."

2. Several biographical books have been written about Parsons. One of the most recent is *Strange Angel: The Otherworldly Life of Rocket Scientist John Whiteside Parsons* by George Pendle (San Diego, CA: Harcourt, 2005).

3. A thorough treatment of this subject may be found in Dr. Zuckerman's book *Behavioral Expression and Biosocial Bases of Sensation Seeking* (Cambridge, UK: Cambridge University Press, 1994).

Chapter 3: Where the Action Is

1. The extract of the Sensation Seeking Scale found in chapter 3 is used with Dr. Zuckerman's permission. The Sensation Seeking Scale has gone through several revisions over time. You can see the original and current test instruments, as well as obtain further information on their interpretation, in *Behavioral Expressions and Biosocial Bases of Sensation Seeking* (Cambridge, UK: Cambridge University Press, 1994).

Chapter 4: Why Live Dangerously?

1. Readers interested in the sociological theory behind edge-working will find the work of the late Dr. Erving Goffman of interest. The University of Chicago–educated Goffman was president of the American Sociological Association and taught for many years at the University of Pennsylvania and the University of California at Berkeley. Highly regarded as a writer, he published several books of tremendous influence.

Goffman's primary methodology was ethnographic study, which basically means he favored observation over statistical data gathering. He was tremendously talented and came to be considered one of the most perceptive "people watchers" in memory.

In a 1967 groundbreaking paper, "Where the Action Is," Goffman wrote of situations in which "fateful actions" occur. Fateful means that a person's next actions could make a significant difference to that individual's entire future.

Goffman wrote that those who take chances and seek opportunities for fateful action build their character. While we intuitively know what is meant when someone is said to have good character or bad character, in "Where the Action Is," Dr. Goffman clearly articulates the attributes upon which a person's character may be judged. Experience gained through taking action in edgy situations builds character, as Goffman defined it. Goffman's article was later published in a collection of his work entitled *Interaction Ritual* (New York: Pantheon Books, 1982).

2. The German study described in chapter 4 was an immense undertaking, and the conclusion relating the relationship between risk and happiness was reported in magazines and newspapers around the world. Falk and his coworkers' study was entitled "Individual Risk Attitudes: New Evidence from a Large, Representative, Experimentally Validated Survey." They analyzed the risk-taking behaviors of a scientifically selected sample of the German population that approximates the characteristics of the country as a whole. The conclusion: "We find a strong positive association between life satisfaction and willingness to take risks in general."

Chapter 7: The Thundering Voice

1. For a highly readable and historical accounting of gunpowder, see Jack Kelley's *Gunpowder* (New York: Basic Books, 2004).

2. Ian von Maltitz's second book, *Black Powder Manufacturing, Testing, and Optimizing* (Dingmans Ferry, PA: American Fireworks News, 2003) is, in my opinion, the best book on the subject of powder making. Of particular interest is his treatment of the selection and manufacture of charcoal.

Chapter 9: The Inner MacGyver

1. Smoke screens are still an important part of commanders' arsenal on the field of battle. According to the U.S. Army Field Manual, FM-3-100/MCWP, "Friendly and enemy surveillance and weapon systems use visual, infrared, or radar sensors to see the battlefield. Smoke and obscurants provide low-cost countermeasures to these systems. Smoke and obscurants can change the

relative combat power of opposing forces by changing the effectiveness of their weapon systems. "

Chapter 10: The Minor Vices

1. On July 1, 2004, the BBC reported: "As the fireworks flare across America on Independence Day, endlessly frustrating The National Fire Protection Association (NFPA), an equally flaming foe will return to the world wide web.

"The website formerly known as zippotricks.com—famous for detailing 555 daring stunts performed with petrol-fuelled Zippo lighters—will mark the Fourth of July holiday by relaunching under a new name.

"The global relaunch, under the name Lightertricks.com, will cause considerable anger, particularly in the U.S. Last summer, the NFPA and U.S. senators, who felt the site was encouraging young people to play with fire, called for Zippotrick's closure. . . . [At about the same time, a horrendous fire during the rock band Great White's pyrotechnic show in a Rhode Island nightclub had sparked an angry outcry.] Comparisons were soon made with the iconic lighter maker Zippo's own nightclub show, which featured professional Zippotricks.com stuntmen showing off their lighter trickery.

"'Federal agencies pressured us to shut it down due to [perceived] hazards with the tricks,' a Zippo spokesperson told BBC News Online.'" ("Website explores dangers of playing with fire," http://news.bbc.co.uk/2/hi/business/3850119.stm, 2004)

2. Wilfred Niels Arnold wrote an article in the June 1989 issue of *Scientific American,* comprehensively describing the history and pharmacology of absinthe. One interesting aspect is the great age of the beverage. Arnold states that "Pliny's Historia Naturalis, written in the first century A.D., describes extracts of wormwood as being of great antiquity (even then!) and having longstanding utility against gastrointestinal worms (hence the name). Thujone

does indeed stun roundworms, which are then expelled by normal peristaltic action of the intestine."

3. During the oil crises of the 1970s, several studies were conducted to determine the effectiveness of the 55 mph speed limit on both fuel conservation and highway safety. Typical of them was "Benefits and Costs of the 55 MPH Speed Limit," published in the *Journal of Policy Analysis and Management*. The upshot was that 55 mph speed limits didn't seem to do much in terms of safety, especially compared to other, more palatable measures.

4. The preceding note should not be interpreted to imply that faster driving is necessarily safer driving. The higher the speed involved in a crash, the more deadly the crash is. See "Velocity Change and Fatality Risk in a Crash—A Rule of Thumb," H. C. Joksch, *Accident Analysis and Prevention,* Vol. 25, No. 1, 1993.

5. The Insurance Institute for Highway Safety lists maximum freeway speeds by state on its Web site, www.iihs.org. This site also lists cell phone use, helmet use, and DUI laws by state as well.

6. The Isle of Man's Tourist Trophy race is an enormously dangerous and extravagantly risky undertaking. Apart from the 100 traffic fatalities occurring in normal driving on the island's roads, an additional 230 Tourist Trophy racers have lost their lives since 1910.

Chapter 11: The Physical Arts

1. Some readers may find whip making as or more interesting than whip cracking. I've made several whips, and while they don't perform as well as those made by professionals, the best hold their own. I recommend *How to Make Whips* by Ron Edwards (Centreville, MD: Cornell Maritime, 1999) as a reference.

2. To learn move advanced knife throwing techniques, see another Harry K. McEvoy book entitled *Knife and Tomahawk Throwing* (North Clarendon, VT: Tuttle, 1973).

3. Arthur Conan Doyle introduces the character of Sherlock Holmes in *A Study in Scarlet*. In chapter 2, Watson describes Holmes as a robust physical figure and notes that Holmes is "an expert singlestick player, boxer, and swordsman."

4. Nearly two dozen Bartitsu self-defense techniques are described and illustrated in *Bartitsu*, written by E. W. Barton Wright and published by *Pearson's Magazine* in January 1901 and February 1901.

Chapter 12: Thrill Eating

1. Gastronomes have long traveled the world in search of unusual and culturally unattractive foods. In 2004, reporter Yaroslav Trofimov explored the strange allure of the Casu Marzu in "Asian Bug Bite: As a Cheese Turns, So Turns This Tale of Many a Maggot—-Full of Larvae, Sardinian Delicacy Flies in the Face of Reason," *Asian Wall Street Journal* (Victoria, Hong Kong: August 24, 2000).

2. On CNN's Web site, CNN Medical Correspondent Elizabeth Cohen reported: "In 1992 the New Jersey Health Department made it illegal to serve undercooked or raw eggs. Violators could be fined $25 to $100. Hysteria resulted. Consumers bemoaned the governmental intrusion into their breakfast tables. Politicians pointed fingers. The law was quickly changed." ("Egg fans uneasy about FDA's 'No over easy' advice," http://archives.cnn.com/2000/FOOD/news/12/08/egg.labeling, 2000)

Chapteer 13: Flamethrowers

1. Readers interested in siege engine projects and history may enjoy my books *The Art of the Catapult* (Chicago: Chicago Review Press, 2003) and *Backyard Ballistics* (Chicago: Chicago Review Press, 2001).

Index